# ANNALS *of* THE NEW YORK ACADEMY OF SCIENCES

**DIRECTOR AND EXECUTIVE EDITOR**
Douglas Braaten

**PROJECT MANAGER**
Steven E. Bohall

The New York Academy of Sciences
7 World Trade Center
250 Greenwich Street, 40th Floor
New York, NY 10007-2157

annals@nyas.org
www.nyas.org/annals

**The New York
Academy of Sciences**

Published by Blackwell Publishing
On behalf of The New York Academy of Sciences

Boston, Massachusetts
2010

# ANNALS *of* THE NEW YORK ACADEMY OF SCIENCES

VOLUME
1211

ISSUE

# Molecular and Integrative Physiology of the Musculoskeletal System

ISSUE EDITORS
Jeffrey I. Mechanick, Li Sun, and Mone Zaidi

## TABLE OF CONTENTS

ANNALS OF THE NEW YORK ACADEMY OF SCIENCES

Issue: *Molecular and Integrative Physiology of the Musculoskeletal System*

# Introduction to *Molecular and Integrative Physiology of the Musculoskeletal System*

Our decision to invite scholarly papers on the topic of molecular and integrative physiology of the musculoskeletal system stems from a long and productive collaborative experience among clinician–scientists at the bench and bedside. Research concepts that were merely conversation pieces at the local café have been realized as the basis for laboratory research and clinical investigation toward novel therapeutic interventions. At first, we planned to devote this issue entirely to bone physiology. After more thought and consideration, however, we decided that including a paper on muscle physiology would create a more provocative and interesting volume, particularly because of the recent interest in anticachexia-targeted therapies. Shortly thereafter, we surmised that this issue was, in fact, more about new paradigms, and as a result we invited papers on arthritis, crossover therapeutics for bisphosphonates in cancer medicine, and bone marrow stem cell research in cardiology. A paper on systems biology and bone pathophysiology was ultimately included in the beginning of the volume to prepare the way for many of the novel concepts to follow. Our personal academic interests have been refreshed by compelling publications by, and discussions with, systems biologists. Moreover, this innovative line of reasoning—to discover new, emergent properties in physiological networks—complements the scientific discoveries of traditional component biology that represent the core of this issue.

The 10 papers in this volume are organized as follows. We first introduce the molecular biology of signal transduction with bone coupling and then present a systems biology overview of how bone is physiologically connected at the organismal level. Then, specialized topics are presented: muscle wasting, arthropathy, and metabolic bone disease in patients with anorexia, spinal cord injury, chronic critical illness, and normal aging. Each of these latter, specialized topics is presented at the molecular and integrative physiology levels and concludes with ideas for future study. The issue concludes with a paper on crossover applications of bisphosphonates into cancer medicine, followed by a paper on bone marrow stem cells and their potential use in heart failure.

There are two main themes discussed in this volume. The first is that the musculoskeletal system does not simply subserve the needs of locomotion and physicomechanical strength. Likewise, it is not solely a reservoir for stored calcium. There is a complex network of molecular signaling both within the bone and also among bone, muscle, and other organs, such as the intestine, brain, pancreas, adipose tissue, and pituitary gland. Owing to these complex pathways, there are many novel molecular targets for drug development to treat disorders that have been heretofore over-looked or neglected. These include cachexia, sarcopenia, osteoporosis, and various deconditioned states associated with an array of chronic illnesses. The second theme is that new paradigms of therapeutic interventions can be realized from a systems approach and from the use of crossover therapies. Indeed, the skeletal system is a hotbed of immunological and hormonal activity, and tapping these resources may provide more effective ways to manage disease. We hope this issue is sufficiently provocative and informative. If these papers challenge the reader to think about

doi: 10.1111/j.1749-6632.2010.05815.x

1

musculoskeletal physiology in new ways and to generate new hypotheses for scientific study, then we have accomplished our goal.

JEFFREY I. MECHANICK, LI SUN, AND MONE ZAIDI
*Mount Sinai School of Medicine*
*New York, New York*

Ann. N.Y. Acad. Sci. ISSN 0077-8923

# ANNALS OF THE NEW YORK ACADEMY OF SCIENCES
Issue: *Molecular and Integrative Physiology of the Musculoskeletal System*

# Cell signaling

Jameel Iqbal,[1,2,3] Mone Zaidi,[3] and Narayan G. Avadhani[1]

[1]Department of Animal Biology, University of Pennsylvania Veterinary School, Philadelphia, Pennsylvania. [2]Hospital of the University of Pennsylvania, Department of Pathology, Philadelphia, Pennsylvania. [3]The Mount Sinai Bone Program, Mount Sinai School of Medicine, New York, New York

Address for correspondence: Jameel Iqbal, M.D., Ph.D., Department of Pathology, Hospital of the University of Pennsylvania, 3400 Spruce Street, Founders Pav. 6th floor, Philadelphia, PA 19104. drjamesiqbal@gmail.com

This review explores advances in our understanding of dynamicism in cellular signaling. Areas highlighted include the role of stochasticity in producing diversity in analogous signaling circumstances; population desynchronization's effect in masking newly appreciated repetitive bursts in protein phosphorylation and messenger RNA production; double-positive feedback interactions and their ability to synchronize multiple signal transduction pathways; scaffolding proteins control over signaling feedback; and frequency-responsive transcriptional regulation as an example of dynamicism in signaling.

Keywords: signal transduction; gene transcription; oscillations; TNF; NF-kappaB; JNK; p38; ERK; temporal; feedback

## Introduction

Signaling experiments demonstrate that many stimuli activate the same signaling pathways, yet result in different outcomes. We recently reviewed the role of complexity in signal transduction.[1] This review is focused on dynamic changes that occur in cell signaling. Among the topics reviewed are stochastic events, population desynchronization, signaling and messenger RNA (mRNA) bursting, mechanisms of feedback control, and frequency-responsive gene induction.

## Stochasticity produces diversity in similar signaling pathways

A single stimulus can lead to diverse outcomes due to random events, as frequently occur in the environment. As humans, we can readily appreciate the random "lucky" or "unlucky" events that help shape our destinies. This randomness or stochasticity plays a crucial role in generating cellular signaling diversity. As an illustration, Lee and colleagues[2] examined why cells of the same type could display two different responses to lipopolysaccharide (LPS) stimulation. They found that LPS induced a

low level of tumor necrosis factor (TNF) production in all cells, but that cells near enough to the other TNF-producing cells had sustained activation of a signaling pathway—nuclear factor kappa B (NF-κB)—that was otherwise transient in nature.[2] For the cells with sustained NF-κB activity, their differential response occurred because of their association or proximity to others (see, Fig. 1). Thus, stochastic events can alter cell fate as the cell integrates these events into its destiny.

Similar to extracellularly driven stochastic events, random intracellular events can encode information about the intracellular environment. However, although extracellularly driven stochastic events depend on being exposed to a ligand at a sufficient concentration, intracellular information is frequently encoded through changes in timing or frequency of a signaling molecule. For example, Cai and colleagues[3] demonstrated that eukaryotic cells encode information in the frequency of stochastic intracellular events rather than the variations in the concentration of signaling molecules. They studied the localization of the Crz1 protein in response to varying concentrations of intracellular $Ca^{2+}$ and found that increasing $Ca^{2+}$ concentrations led to

doi: 10.1111/j.1749-6632.2010.05811.x

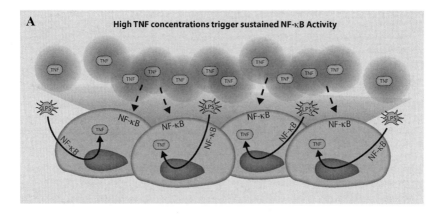

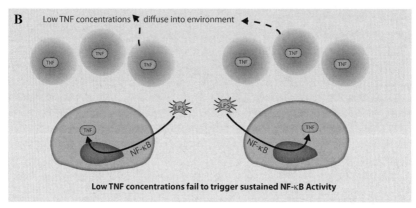

**Figure 1.** Stochastic extracellular events can result in different signaling outcomes. Lipopolysaccharide (LPS) signaling can result in either sustained or transient nuclear factor kappa-light-chain-enhancer of activated B cell (NF-κB) activity, depending on whether the cells are exposed to a sufficient paracrine tumor necrosis factor (TNF) signal. (A) LPS-induced TNF production by cells in close proximity can raise TNF concentrations to trigger a second wave of NF-κB signaling, thereby resulting in persistent NF-κB activity. (B) Cells not in close proximity may not be exposed to TNF in sufficient quantity to activate TNF-induced NF-κB signaling; the result is that one stimulus, such as LPS, can result in two distinct outcomes based on environmentally controlled variables.

a greater frequency of nuclear translocation and bursts of mRNA production, but without any significant changes in amplitude or duration of activation.[3]

Intracellular signaling has evolved to depend more on frequency modulation than amplitude modulation to encode information, probably because cells vary greatly in size and protein amounts. For instance, Cohen-Saidon and colleagues[4] studied extracellular signal-regulated kinases 2 (ERK2) activation and showed that concentration parameters, such as basal levels and peak heights of the ERK2 response, are variable from cell-to-cell, but that the timing parameters are relatively consistent and encoded the signaling information. Thus, intracellular signal transduction is commonly de-pendent on timing and not concentration in order to improve the robustness of information encoding.

## Static-appearing phosphorylations are an averaging artifact

Stochastic changes within a population create subtle differences between cells that over time lead to desynchronization. Although rapid phosphorylation of signaling molecules has been noted for decades, until recently, it was assumed that the phosphorylation of signaling pathways at later time points was gradual in nature and sustained; that is to say, rapid, pulse-like behavior was not characteristic of signaling events that occur hours after a stimulus. It is now clear, however, that many of these previous observations were confounded by population desynchronization. On an individual cell basis, the rapidity and brevity of signal transduction is often preserved; for example, this preservation is observed as reoccurring phosphorylation bursts.[5,6] Thus in addition to determining which pathways are activated for a given ligand, we now appreciate that the repetitive activation of these pathways is used to encode a diverse set of information.

Similar to kinase-mediated protein phosphorylation, a paradigm shift in our understanding of the kinetics of mRNA production has also occurred—it is now appreciated that signaling-induced mRNA production is a highly dynamic event that results in rapid rises and falls in mRNA levels over a matter of minutes.[7] mRNA production consists of rounds of transcriptional activation taking place over minutes, with multiple mRNAs produced per cycle.[8] Interestingly, the transcriptional factors themselves appear to bind only seconds along each stage of the "transcriptional clock cycle."[9] Indeed, it is possible that a common gene induction mechanism in eukaryotes relies on a 60 min transcriptional clock, which is temporally constrained by covalent histone modifications and the kinetics of RNA polymerase activity.[8] However, not all investigators support this notion, and at least for the glucocorticoid receptor, it has been demonstrated that transcriptional bursting does not always occur.[10]

A desynchronization effect in mRNA production is frequently observed after the completion of the first cycle, because some cells have finished earlier than others; thus, the duration of subsequent cycles becomes progressively broader in distribution over time.[8] In other words, the noise in cycle duration increases linearly with the number of cycles completed.[8] Nonetheless, we have demonstrated using time-series microarrays and quantitative PCR that hundreds to thousands of genes appear to conform to the 60 min "transcriptional clock" when a stimulus is applied (see Fig. 2).[7] Similarly, glucocorticoid-induced mRNA changes have been shown to be highly dynamic, with induction and diminution occurring in less than 60 min.[10]

## Positive feedback loops synchronize multiple signaling pathways

Most signal transduction events activate multiple signaling pathways. Given that the activation frequency of intracellular signaling pathways is commonly used to encode information, a burgeoning understanding is occurring regarding the temporal coordination of multiple signaling pathways. Feedback mechanisms can produce oscillatory dynamicism in most signaling pathways. Computer modeling has recently demonstrated that two independent oscillatory signaling pathways can be synchronized if they share a weak interaction.[11] This weak interaction may manifest as a small intracellular change in a shared component between pathways.[11] For example, synchronizations of rises in cyclic adenosine monophosphate (cAMP) and $Ca^{2+}$ are required for insulin secretion from pancreatic cells, and this synchronization is dependent upon mutual protein kinase A activation within each pathway.[12] Similarly, there has been one demonstration of synchronization of multiple kinase signaling pathways over time (see, Fig. 2); human T cells exposed to TNF display repetitive coordinated bursts of NF-κB, c-Jun N-terminal kinases (JNK), and p38 mitogen-activated protein kinase (MAPK) activation, approximately every 2 hours.[5]

What type of feedback mechanism is required to produce the observed synchrony of signaling molecule activation? Computer modeling suggests that double-positive feedback coupling can explain how many pathways are synchronized when only weakly coupled to an environmentally driven master oscillator such as NF-κB.[11] One finding from computer simulations regarding double-positive feedback coupling is that large changes in the amplitude and periodicity of signaling pathway activation can occur by varying the coupling strength.[11] Using the NF-κB pathway as an example, we were able to confirm that both the amplitude and periodicity of signaling molecule activation can dramatically change when altering one of the coupled pathways.[5] Through intracellular phospho-flow cytometry, we demonstrated that inhibition of NF-κB activation through expression of a dominant-negative IκBα changed the frequency and amplitude of both JNK and ERK activation.[5]

Most experimental evidence indicates that changes in the frequency of signal transduction pathway activation are more important than changes in amplitude; however, an almost universal finding is that at later time points, the strength of signaling pathway activation is weaker. An explanation for this phenomenon is that there is incomplete resetting of the individual intracellular

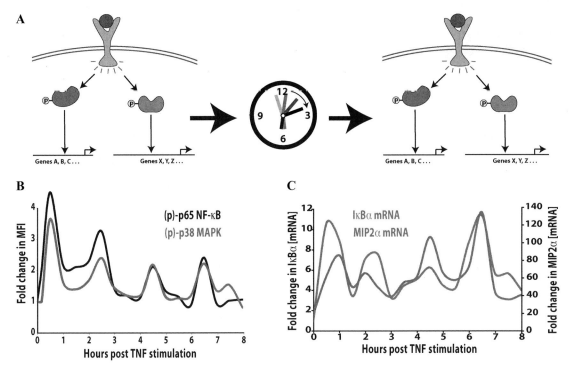

**Figure 2.** Signal transduction and gene expression can occur in short, repetitive bursts. (A) A schematic displaying repetitive stimulation of two signaling pathways that lead to gene induction is shown. (B) Experimental data from Jurkat human T cells stimulated with TNF demonstrating coordinated, repetitive burst-like activations of (p)-JNK1/2 and (p)-p65 NF-κB as analyzed by flow-cytometry (median fluorescence intensity, MFI). (C) A time course of mRNA production in primary murine macrophages demonstrating coordinated, repetitive gene transcription in the IκBα and macrophage inflammatory protein2α genes following TNF stimulation. Figure 2(B) was adapted from Iqbal and Zaidi,[5] and Fig. 2(C) was adapted from Sun *et al.*[7]

signaling components.[13] For example, decreased responses in TNF-induced NF-κB translocation are caused by quantitative differences in the amount of phosphorylated IκBα, which through wash out and repetitive stimulation experiments was shown to result from a failure to completely reset the signal transduction machinery. Overall, signaling pathways are highly dynamic, and double-positive feedback loops influence their synchronicity.

## Scaffolding proteins can control stimulation dynamicism

What mechanisms do cells use to control the frequency of activation and double-positive feedback synchronization? Scaffold proteins represent a common mechanism used to control spatial and temporal aspects of signaling. For example, epidermal growth factor (EGF)-induced Ras activation results in different outcomes depending on the scaffold proteins used to induce Ras activation; at different intracellular sites, Ras induction can either result

in feedback inhibition of the EGF receptor or in arachidonic acid production.[14] Likewise, differential feedback wiring of the ERK cascade can result in distinct temporal profiles of activation, as occurs for the ligands EGF and nerve growth factor (NGF).[14] Scaffold proteins can structure the nature of the feedback; using the ERK cascade as an illustration, EGF was shown to elicit negative feedback, whereas NGF induced positive feedback.[14] Thus, scaffold proteins can alter the nature of the feedback, thereby dramatically altering the kinetics of activation.

Scaffold-mediated and nonscaffold signaling can occur simultaneously. Blocking a protein in one pathway in the scaffold or free signaling pathway can swing the signaling entirely onto the other pathway and result in altered kinetics and feedback. For example, G protein-coupled receptor-induced ERK activation displays transient nonscaffold activation and β-arrestin-mediated sustained activation[14]; artificially blocking either pathway results in signaling only through the nonblocked mechanism.[14] Thus, scaffolds can be used to add another layer of specificity and frequency for signal transduction.

## Variations in stimulation frequency encode unique outcomes

The purpose of dynamicism in both signaling pathway activation and gene induction is to allow for distinct outcomes from different activation frequencies. For example, varying the frequencies of NF-κB nuclear entry leads to differential regulation of particular downstream genes.[13] Ashall and colleagues demonstrated that for low frequencies of NF-κB activation, there was no production of some genes, e.g., *RANTES*, which were instead induced only at higher frequencies of NF-κB activation; in contrast, they showed that other genes, such as *IκBα*, were induced regardless of the frequency of activation.[13] Thus, different frequencies of signal pathway activation can lead to unique transcriptional outcomes. Similarly, pulses of glucocorticoid activity result in a different genetic output compared with continuous glucocorticoid activity.[15]

Many cellular outcomes depend on multiple signaling pathways being simultaneously activated, and the dynamic nature of signaling pathway activation allows combinatorial induction to occur. For instance, $Ca^{2+}$-induced exocytosis of insulin requires cAMP-mediated amplification.[12] Similarly, some NF-κB-dependent genes require p38 MAPK-mediated histone promoter processing for their induction, with disruption of p38 MAPK blocking transcription despite NF-κB activation.[16] In line with this, it has been experimentally demonstrated that reoccurring bursts of p38MAPK activation are temporally aligned with NF-κB activation.[5] Some genes may require more than just two pathways for induction such as interleukin 6 (IL-6) with its enhanceosome activated by NF-κB, p38 MAPK, JNK1/2, and ERK1/2 signaling.[17] In an analogous manner to the previous examples, correlation of the temporal alignment of all four signaling pathways with IL-6 induction has been demonstrated.[7]

## Conclusions

This review has described advances in our understanding of cellular signaling. Research demonstrating the importance of stochastic events has revealed its role in producing diversity in seemingly analogous signaling circumstances. A shift in our existing thinking regarding stimulus-induced protein phosphorylation and mRNA production was espoused, where it is now appreciated that population desynchronization masks repetitive bursts in protein phosphorylation and mRNA production that are occurring over time. New mathematical models have revealed that repetitive activations in multiple signal transduction pathways can be synchronized through double-positive feedback interactions. The way cells commonly control signaling feedback is through scaffolding proteins. Lastly, the meaning of dynamicism in signaling was explored, and research demonstrating the ability of transcriptional regulation to be frequency responsive was reviewed.

## Acknowledgments

J.I. wishes to acknowledge past support from the AFAR. Drs. Mone Zaidi and Narayan Avadhani are supported by grants from the U.S. National Institutes of Health.

## Conflicts of interest

Dr. Zaidi consults for Genentech, Amgen, and Warner Chilcott. Dr. Zaidi is also a named inventor of a pending patent application related to osteoclastic bone resorption filed by the Mount Sinai School of Medicine (MSSM). In the event the pending or issued patent is licensed, he would be entitled to a share of any proceeds MSSM receives from the licensee. No other potential conflicts of interest relevant to this article were reported.

## References

1. Iqbal, J., L. Sun & M. Zaidi. 2010. Complexity in signal transduction. *Ann. N.Y. Acad. Sci.* **1192:** 238–244.
2. Lee, T.K. *et al.* 2009. A noisy paracrine signal determines the cellular NF-kappaB response to lipopolysaccharide. *Sci. Signal* **2:** ra65.
3. Cai, L., C.K. Dalal & M.B. Elowitz. 2008. Frequency-modulated nuclear localization bursts coordinate gene regulation. *Nature* **455:** 485–490.
4. Cohen-Saidon C. *et al.* 2009. Dynamics and variability of ERK2 response to EGF in individual living cells. *Mol. Cell* **36:** 885–893.
5. Iqbal, J. & M. Zaidi. 2008. TNF-induced MAP kinase activation oscillates in time. *Biochem. Biophys. Res. Commun.* **371:** 906–911.
6. Nelson, D. E. *et al.* 2004. Oscillations in NF-kappaB signaling control the dynamics of gene expression. *Science* **306:** 704–708.

7. Sun, L. *et al.* 2008. TNF-induced gene expression oscillates in time. *Biochem. Biophys. Res. Commun.* **371:** 900–905.

8. Degenhardt, T. *et al.* 2009. Population-level transcription cycles derive from stochastic timing of single-cell transcription. *Cell* **138:** 489–501.

9. Hager, G.L., J.G. McNally & T. Misteli. 2009. Transcription dynamics. *Mol. Cell* **35:** 741–753.

10. Stavreva, D.A. *et al.* 2009. Ultradian hormone stimulation induces glucocorticoid receptor-mediated pulses of gene transcription. *Nat. Cell Biol.* **11:** 1093–1102.

11. Kim, J.R. *et al.* 2010. A design principle underlying the synchronization of oscillations in cellular systems. *J. Cell Sci.* **123:** 537–543.

12. Idevall-Hagren, O. *et al.* 2010. cAMP mediators of pulsatile insulin secretion from glucose-stimulated single beta-cells. *J. Biol. Chem.*

13. Ashall, L. *et al.* 2009. Pulsatile stimulation determines timing and specificity of NF-kappaB-dependent transcription. *Science* **324:** 242–246.

14. Kholodenko, B.N., J.F. Hancock & W. Kolch. 2010. Signalling ballet in space and time. *Nat. Rev. Mol. Cell Biol.* **11:** 414–426.

15. Desvergne, B. & C. Heligon. 2009. Steroid hormone pulsing drives cyclic gene expression. *Nat. Cell Biol.* **11:** 1051–1053.

16. Saccani, S., S. Pantano & G. Natoli. 2002. p38-Dependent marking of inflammatory genes for increased NF-kappa B recruitment. *Nat. Immunol.* **3:** 69–75.

17. Lu, H. *et al.* 2005. Regulation of interleukin-6 promoter activation in gastric epithelial cells infected with *Helicobacter pylori*. *Mol. Biol. Cell* **16:** 4954–4966.

Ann. N.Y. Acad. Sci. ISSN 0077-8923

ANNALS OF THE NEW YORK ACADEMY OF SCIENCES

Issue: *Molecular and Integrative Physiology of the Musculoskeletal System*

# A systems approach to bone pathophysiology

Aaron J. Weiss,[1] Azi Lipshtat,[2] and Jeffrey I. Mechanick[1]

[1]Division of Endocrinology, Diabetes, and Bone Disease, Mount Sinai School of Medicine, New York, New York. [2]The Gonda (Goldschmied) Multidisciplinary Brain Research Center, Bar-Ilan University, Ramat-Gan, Israel

Address for correspondence: Jeffrey I. Mechanick, M.D., Division of Endocrinology, Diabetes, and Bone Disease, Mount Sinai School of Medicine, 1192 Park Avenue, New York, NY 10128. jmechanick@aol.com

With evolving interest in multiscalar biological systems one could assume that reductionist approaches may not fully describe biological complexity. Instead, tools such as mathematical modeling, network analysis, and other multiplexed clinical- and research-oriented tests enable rapid analyses of high-throughput data parsed at the genomic, proteomic, metabolomic, and physiomic levels. A physiomic-level approach allows for recursive horizontal and vertical integration of subsystem coupling across and within spatiotemporal scales. Additionally, this methodology recognizes previously ignored subsystems and the strong, nonintuitively obvious and indirect connections among physiological events that potentially account for the uncertainties in medicine. In this review, we flip the reductionist research paradigm and review the concept of systems biology and its applications to bone pathophysiology. Specifically, a bone-centric physiome model is presented that incorporates systemic-level processes with their respective therapeutic implications.

Keywords: metabolic bone; physiome; systems biology; skeletal system; vitamin D

## Introduction

Over the past century, substantial progress has been made in understanding the functions of the human body at the molecular, physiological, and systemic levels. Scientific research has advanced from the discovery of the basic DNA structure to a fully characterized human genome. Each piece of information provides another valuable tool to help guide the ever-evolving practice of medicine. Yet, the more information that is uncovered, the more complexity is realized and unsolved. For example, Watson and Crick's description of the structure of DNA was paramount and provided a basis for future discoveries; however, this basic architecture was just a stepping stone in the uncovering of epigenetic markers, structural mutations, and single nucleotide polymorphisms (SNPs) that better explain the function and regulation of DNA. Presently, international collaborations among large-scale sequencing centers specializing in generating and analyzing high-throughput data are assessing human genetic variation and how genomic characterization can be applied toward personalized medicine.[1] The contin-ued push for greater and better ways to understand the biological milieu has shaped our current understanding of the human body.

Inductive reasoning and hypothesis-driven experimentation are the driving forces behind reductionism ("bottom-up" research) in medical science. In order to understand nature, reductionist scientists break down the proposed system into its component parts so as to tease out individual functions. In short, this method is summarized by the system equaling the sum of the parts and understanding the system as a whole depends on defining the important interactions of each of the system's constituents. A reductionist's approach to bone pathophysiology would entail, for example, breaking bone down into its component parts—the osteoblast, osteocyte, and osteoclast—and then determining the individual cell's function and the interactions it has with its neighbors. This would yield information that can be applied to produce a larger system of bone remodeling.

In the biological world, reductionism began with the developmental concept of a gene by Wilhelm Ludvig Johannsen in the early 20th century. The

doi: 10.1111/j.1749-6632.2010.05816.x

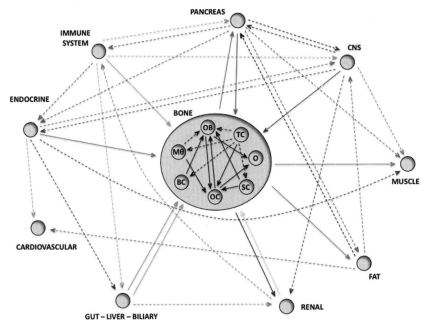

**Figure 1.** Theoretical construct of some integrative bone physiology connections. Indirect connections on bone are illustrated with dashed arrows; direct connections are marked with solid arrows. Abbreviations: OB = osteoblast; TC = T-lymphocyte; O = osteocyte; SC = stromal cell; OC = osteoclast; BC = B-lymphocyte; Mθ = macrophage; CNS = central nervous system; SNS = sympathetic nervous system; OCN = osteocalcin; PTH = parathyroid hormone; CT = calcitonin; T3/T4 = thyroid hormones; 1,25-D = 1,25-dihydroxyvitamin D; Ca = calcium; Phos = phosphorus; FGF-23 = fibroblast growth factor-23; MCH = melanin concentrating hormone.

belief at the time, and for many years after, was that a single gene passed on a single trait through generations. Although at the time, the idea of DNA sequences as being the true courier of heritability was unknown, the foundation had been laid for the "one gene, one protein" hypothesis. This scientific dogma peaked with the awarding of the Nobel Prize in Physiology or Medicine to George Wells Beadle and Edward Lawrie Tatum for their discovery of genes regulating cellular biochemical events. For years, researchers spent unimaginable amounts of time, money, and effort into disassembling biology to its component parts in order to unveil direct and discrete relationships between "genotype" and "phenotype."

By itself, the reductionist approach proved effective when, due to the lack of technology, our understanding of disease was limited to one specific genotype leading to one specific phenotype. An example of this is the monogenic disease involving the cystic fibrosis transmembrane conductance regulator (CFTR) gene mutation. Yet, even this classic explanation of a common disease has proven to be oversimplified as further research has shown there to be over 1,400 different disease-causing mutations that correspond to unpredictable levels of phenotypic expression.[2] Further championing this methodology was the development of gene knockout models in living systems. However, deletion of genetic information through knockout models often does not change the phenotype as one might expect. Instead, it is the system's natural robustness— defined as the ability of a complex biological system to maintain its function despite internal and external perturbations—that often allows for retention of the initial trait.[3] Robustness is a fundamental property of systems-level nonlinear phenomena that cannot be understood through reductionism.[3]

Nevertheless, when a critical amount of stress is applied to a biological system, the deleted genotype often manifests the expected new phenotype. Hillenmeyer and colleagues[4] found that deleting up to 80% of genes in yeast had no phenotypic consequence. After stressing the yeast (in this experiment the yeast is the system) they found that up to 97% of the deleted genes resulted in a measurable growth

**Table 1.** High throughput tools used for systems biology approaches

| Tool | Definition | Examples |
|---|---|---|
| Two-hybrid screening | Technique used to discover physical interactions (interactome) between two molecules in a biological system. | Protein–protein interactions, protein-DNA interactions |
| Text mining | Process of deriving high-quality, high-throughput information from text. | PubGene, GoPubMed.org |
| Genomewide scans | Assays that measure hundreds of thousands of points of variation in a person's genome simultaneously | |
| Genome wide association studies | Examination of the genetic variability within individuals of a specific group and subsequent association of the variation with particular traits or diseases | Klein *et al.*,[170] Sladek *et al.*,[171] Wellcome Trust Case Control Consortium[172] |
| Microarrays | An array that assays biological material using high-throughput screening methods to determine the presence and amount of a particular entity. | cDNA, SNP, protein, MMChips, tissue, cellular, and antibody arrays |
| Tandem mass spectrometric analysis | A sequencing technique commonly used to identify structure of compounds. Commonly used to sequence the amino acids in a specific protein. | Phosphoproteomics |
| Next-generation sequencing | DNA sequencing of multiple areas of the genome rapidly and simultaneously at a reduced cost. | Chipseek, mRNAseek |
| Curated databases | Large scale collaborations that seek to develop centralized data storage vehicles focusing on a specific scale or system. The databases allow for high-throughput data analysis of complex systems. | OMIM, HapMap, Human Genome Project, Physiome Project, Virtual Physiological Human Project, DrugBank |
| Mathematical modeling | Describing a system (genomic, cellular, biological, sociological, etc.) using mathematical language. Advantage is that systems from various contexts and fields can be similarly analyzed and compared. | Ordinary differential equations, partial differential equations, network models, Boolean networks, Bayesian networks |

phenotype, suggesting additional scales of complexity contributing to phenotypic expression.[4] In addition, control of protein-coding DNA by nonprotein coding DNA, splice variants, environmental-induced genome rearrangements, nonrandom mutations, gene transfer, SNPs, linkage disequilibrium, and epigenomics all disrupt the ideal "one gene, one protein" concept and demonstrate the inherent multiscalar complexity of biological systems.[5] Clinical application of reductionist-derived data may account for the

phenomenon where incomplete knowledge gaps can lead to erroneous diagnoses and therapeutic interventions. Closing these knowledge gaps will optimize decision making and patient care.

One potential technique to address the problem of incomplete knowledge is based on the premise that failure to recognize strong, nonintuitively obvious, and indirect connections among physiological events accounts for uncertainties. Applying a whole-body physiomic network and recognizing the importance of previously ignored subsystems, such as the skeletal subsystem, can provide valuable information to help clinical decision making as it relates to various disease treatment paradigms. Although classically known to provide the functions of locomotion and mineral homeostasis, the skeletal system interacts and controls whole body physiology in an even more labyrinthine manner (Fig. 1). The basic functioning unit of the skeletal system, the osteoblast–osteoclast coupling model shows remarkable complexity. Various paracrine signals—cytokines, neurotransmitters, and other molecular effectors—work directly and indirectly to achieve homeostatic control over osteoblastic bone formation and osteoclastic bone resorption, which characterize the skeletal microphysiological environment. Trans-system events involving bone are maintained on the macrophysiological level by hormonal, neuronal, and physico-mechanical signals, such as the newly discovered connections with brain, pancreas, and gut.[6–12] It is this recursive horizontal and vertical integration of subsystem coupling across and within spatio-temporal scales that perpetuates living tissue.

However, despite a core literature on bone and its connections with other organ systems, bone is often ignored in the setting of complex systemic dysfunction such as that seen in critical illness, cancer, and cardiovascular (CV) disease. In the sections that follow, we first step back and introduce the concept of systems biology and the application of small-world networks to biological systems such as the skeletal system. Then, a discussion of other emerging bone-centric physiome models spanning specific pathophysiological disease phenomenon with their respective therapeutic implications is presented.

## Systems biology

The increased understanding of the complexity of multiscalar biological systems and the development of multiplexed tests and computer technology that can rapidly generate and analyze terabytes of information provides exciting opportunities never before thought possible (Table 1). One result of this technological explosion is the growth of a novel field called "discovery science," a scientific methodology that derives new findings and hypotheses from experimental analysis of large volumes in an effort to define the components of a system.[13] Applications of discovery science include the Human Genome Project and later, at higher levels of scale, the Phys-

iome Project[14] and Virtual Physiological Human Project.[15,16] Combining both the hypothesis driven "bottom-up" experimentation with a discovery science ideology paved the way for the evolution of systems biology, a term used to describe the behavior and relationships of all of the elements of a specified functioning biological system.[13]

One systems biology approach is based on graph theory methods that visualize the system as a network and then attempt to analyze the component nodes (parts) and edges (connections or interactions) and what they mean in terms of functionality. Systems biology incorporates all of the components of each hierarchical level and seeks to integrate the interactions at each scale (horizontal integration) and among different scales (vertical integration) to develop a holistic model (Fig. 2). Systems biology focuses on a "top–down" approach that provides a more thorough understanding of the topology of structural relationships and dynamic interactions that characterize large, complex systems. Analysis of the system yields previously unknown relationships that are known as "emergent motifs," a result not possible under traditional reductionist strategies. Although the exact details may differ, the above approach involves four steps:[17]

Step 1: define the system;

Step 2: identify the nodes of the system;

Step 3: determine the edges between the nodes of the system; and

Step 4: model the spatiotemporal dynamics of the network mathematically.

After steps 1–4 have been completed, confirmation of the emergent motifs through scientific experimentation may be performed to validate the findings. Thus, true systems biology incorporates both reductionism and holism in its attempt to explain how component parts interact to allow for phenotypic expression.

One notable advantage to this method is that systems from various contexts and fields can be similarly analyzed, allowing comparison of their specific overall structure, individual connections, and emergent properties. For example, in a sociological system, each node represents one individual, and in a biological system, a node may represent a protein in an osteoblast or a specific species in an ecological system, depending on the defined scale. To continue the example from above, the individuals (nodes) may relate as co-workers (edge) in the

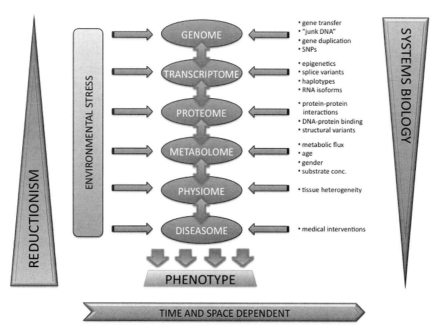

**Figure 2.** Multiple scales of biological activity. With the evolving interest in biological systems, classical reductionist approaches may not fully describe their complexity. The reductionist approach, left, breaks the system down into its component parts and individually parses the effects that each of the components has on each other. A systems biology approach, right, takes into account all of the components and interactions at each scale, vertically integrates the multiple scales, and elicits the spatiotemporal dynamics of the system to provide an overarching, holistic characterization of the phenotype. DNA = deoxyribonucleic acid; RNA = ribonucleic acid; SNPs = single nucleotide polymorphisms.

sociological system, and the two proteins (nodes) may interact enzymatically (edge) in the biological system. Other examples of large-scale networks include scientific collaboration networks where authors of scientific manuscripts are connected if they have at least one joint publication;[18] in cell biology, protein–protein interaction networks have been constructed and analyzed for various organisms;[19,20] and networks that connect drugs to their targets, diseases, or side effects are used for predicting innovative drug designs.[21,22]

For systems biology studies, complex systems analysis techniques, such as mathematical modeling and network analysis, are well suited to integrate the high throughput data necessary to define the system's architecture and produce a simplified graphic representation that can be interrogated for nonintuitive properties. Complex systems analysis has been applied to many scientific disciplines including social networking,[23–25] anthropology,[26] economics,[27] and physics.[28,29] In addition, these techniques have been used on the level of the genome,[30,31] pro-

teome,[32–35] and metabolome.[36] At the level of the physiome and diseasome, complex systems analysis will incorporate individual genes, proteins, metabolites, reactions to stress, and environmental influences in order to explain various levels of phenotypic expression at a macroscopic level.[37] In turn, the knowledge gained from this overarching view will potentially enable the development of specific interventions to modify and prevent disease.

With the current state-of-the-art of available genomic, transcriptomic, proteomic, metabolomic, and even physiomic datasets, we are at an exciting point where network analysis of complex biological systems is a reality.[38] As stated earlier, a system is composed of nodes. In network analysis, the most fundamental property of a node is its degree, or connectivity, which is the number of edges connected to the node. High connectivity indicates that the node is important and that removing that node may significantly alter the overall structure of the system. The structure of a system is characterized by the distribution of the node degrees. Different systems

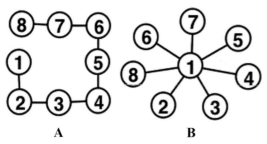

**Figure 3.** Comparison of system structures. Both of the graphs (systems) above, A and B, have a set of eight nodes and seven edges. Therefore, the average degree is 1.75 in both of them. However, the structure differs significantly, as A has an average distance between nodes of three edges, whereas B has an average distance between nodes of 1.75 edges.

may have similar average degrees, but very different structures (Fig. 3). The distribution of node degrees captures this overall structure and allows comparison among different systems.

Due to growth and preferential attachment, the topology of most complex biological networks is characterized as being "scale-free."[38] Scale-free describes a pattern where certain nodes are more highly connected than others, and the degree distribution can be described by a "power law."[39,40] Power laws suggest self-organization within a system. Scale-free systems are characterized by a few large "hubs" (nodes with very high connectivity) and thus have multiple paths connecting any two nodes. Thus, removal of random nodes from the system will not affect its overall connectivity as long as the hubs remain intact. This type of system will avoid uncoupling even with the removal of a relatively large fraction of randomly selected nodes.[41]

The average distance between nodes (i.e., the length of the shortest path between them) is an important measure of the overall network. Many real life networks exhibit a low mean distance; these networks are known as "small-world" networks.[42] Small-world networks lack direct connections between most pairs of nodes but instead connect them through small indirect pathways. This small-world effect has held true in systems studies ranging from neural networks to the World Wide Web.[38] The most famous example is the "six degrees of separation" idea that claims that any two people in the world can be connected by a friendship chain (a friend of a friend, etc.) that has six steps or less. In-

creasing the distance between many pairs of nodes will in turn have a significant effect on the entire network activity. In addition, the short path length that characterizes small-world networks allows local perturbations to spread quickly throughout the network.[38] This was first noted in metabolism studies where only three to four steps (or reactions) link most metabolites.[36,43]

Beyond local and statistical analyses of the network's topology, much interest is given to the spatiotemporal dynamics of the network. For example, in cell signaling systems, not all of the reactions take place simultaneously. Furthermore, any intracellular reaction affects the concentration of the respective reactants and thus changes their availability for other potential reactions. Given a set of initial concentrations one can follow (analytically or by numerical simulations) the time course of the biochemical activity in the network. These dynamics emphasize the role of indirect connections and specific topological modules and serve as a tool for evaluating the importance of particular nodes or edges.

Despite the random nature of real-world networks, many networks have over representation of small modules, termed network motifs.[44] These motifs are small subgraphs (usually consisting of three to four nodes) that play a large role in determining the dynamics of the network. Although still evolving, the methodology to quantitate the dynamics of biological networks includes reverse engineering,[45] nonlinear differential equations with multiple linear regression,[46] bifurcation analysis,[47] Bayesian analysis,[48] Boolean networks,[47] and cellular automata.[49]

The physiome vertically integrates all of the -omics of lower molecular scales with organismal macrophysiological scales, providing a hierarchy within the defined system. This systems biology approach has intriguing implications. First, the success of mathematical modeling and network analysis at the molecular scale provides exciting hope that the same success may occur at the macrophysiological scale. Second, translational applications of physiomic emergent motifs may be realized once validated by conventional bottom-top clinical studies. Third, holistic systems biology approaches toward the physiome can potentially elucidate the structure of previously ignored components of specified subsystems.

# Bone-centric physiomics

## General statements

Interrogating a bone-centric physiome from a global perspective can provide useful insights about strong, nonintuitive, indirect connections, and it is the vast complexity innate to the skeletal system that makes it a prime candidate to test the applicability of this holistic method. In addition, the skeletal system interacts with other physiological systems, like the renal, CV, immune, and endocrine systems (Fig. 1). The hierarchical structure of this bone-centric physiome potentially enables analysis within and between systems. A comprehensive analysis could isolate the relevant components within the skeletal system and provide a more thorough understanding of the mechanism by which bone characteristics can affect other systems and disease states.

## Osteoporosis

Low bone mass and increased fracture risk characterize the complex progressive disease of osteoporosis.[50] Many types of osteoporosis (postmenopausal, senile, medication induced, etc.) exist, and the correct therapy depends upon identifying the specific etiology. Often, however, the complexity of multiple contributing factors makes a precise identification of a single cause impossible. The complex interplay of multiscalar level events within bone as well as characterization of the between-subsystem interactions involving bone can characterize the biological basis of osteoporosis on a personalized level.

Important to the underlying structure of bone in both health and disease is the osteocyte. The osteocyte's ability to respond to physico-mechanical stress and strain as well as humoral factors allows for the initiation of bone remodeling and formation, thus making it a unique possibility to explore further the underlying genetic interactions working to maintain skeletal homeostasis. Recent studies done on the osteocyte propose this cell to be even more dynamic and interactive than previously thought.[51,52] Dean and colleagues[53] used microarrays to identify 269 genes overexpressed in osteocytes of which many have previously known roles in bone and muscle differentiation and contractility. With this data, they determined the evolutionarily conserved and enriched transcription factor binding sites in the promoter regions of these genes, which then allowed them to construct an osteocyte transcriptional regulatory network. Additionally, Paic

and colleagues[54] isolated osteocyte and osteoblast cell populations and elicited the differential gene expression by comprehensive microarray analysis of their gene profiles. They identified a number of osteocyte-expressed genes previously known to be important for muscle and neuronal development and function. Further understanding of the complex interactions at the genomic level within the osteocyte and other cellular constituents of bone as well as mapping of the proteomic and metabolomic scales may help aid in defining interventions to prevent osteoporosis.

Novel techniques exploring the complex interactions within the bone genome are being developed to exploit the scattered base of knowledge present on the Internet. Gajendran and colleagues[55] created a text-mining tool that farmed the PubMed literature to elicit a detailed map of the genes and interrelationships within the osteoporosis network. They identified previously unstudied genes with respect to osteoporosis that have the potential for therapeutic intervention at the genomic level. As shown in this study, tools such as text-mining prove incredibly valuable in the struggle to incorporate the genomic, proteomic, and physiomic level interactions in both health and disease.

Although treatment for individual patients must take into account the patient's age, risk-factors, and generalized medical conditions, the goals of osteoporosis therapy are the same and include prevention of decreased bone mass and a reduction in fractures. In addition to lifestyle interventions that include smoking cessation, increased physical activity, and fall prevention, determining a patient's vitamin D status with supplementary treatment is now more than ever part of routine care for patients with suspected osteoporosis. The typical first-line pharmacological interventions for osteoporosis include 1,000–1,500 mg/day of calcium and 400–1,000 units/day of vitamin D depending on age and other demographic and clinical factors.[56] Not only will identifying a patient's vitamin D status allow for optimal osteoporosis management, but it will also help prevent the development of other deleterious nonskeletal health effects that result from vitamin D deficiency (Table 2).

Most agents used to prevent further progression of osteoporosis are either antiresorptive medications, which decrease bone resorption, or anabolic medications, which promote bone formation.

**Table 2.** Causes and consequences of vitamin D deficiency

| Causes | Potential health consequences |
| --- | --- |
| Aging | Depression |
| Dietary | Schizophrenia |
| Inadequate sunlight | Infectious disease |
| Obesity | Decreased forced |
| Malabsorption (cystic | expiratory volume in |
| fibrosis, Crohn's | one second |
| disease, Celiac disease, | Asthma |
| Whipple's disease) | Congestive heart failure |
| Postgastrectomy | Hypertension |
| Pancreatic Insufficiency | Type 2 diabetes |
| Hepatic insufficiency | Syndrome X |
| Medications | Autoimmune diseases |
| (anticonvulsants, | (type 1 diabetes, |
| steroids, rifampin, | multiple sclerosis, |
| highly active | Crohn's disease, |
| antiretroviral therapy) | rheumatoid arthritis) |
| Nephrotic Syndrome | Muscle weakness and |
| Hypoparathyroidism | aches |
| Renal Failure | Osteoarthritis |
| Vitamin D-dependent | Osteoporosis |
| rickets | Osteomalacia |
| | Rickets |
| | Cancer (breast, prostate, |
| | colon, pancreas, acute |
| | myeloid leukemia, |
| | myelodysplastic |
| | syndromes) |

Currently available interventions outside the immediate skeletal system but have proven benefit include gonadal steroids and receptor-activator of NFκB-ligand (RANKL) inhibitors. Denosumab, an anti-RANKL humanized monoclonal antibody, acts by inhibition of osteoclast-mediated bone resorption. Its effects on postmenopausal osteoporotic women were studied in the Fracture Reduction Evaluation of Denosumab in Osteoporosis Every 6 Months (FREEDOM)[57] and Study of Transitioning from Alendronate to Denosumab (STAND)[58] studies, both of which showed reduced bone turnover and fractures in the treatment groups.[59] Although further studies are needed, physiomic investigations identifying potential biological agents, small molecules, or other pharmaceuticals with known

activity in nonskeletal tissue that are not intuitively thought to act on bone have proven useful in combating osteoporosis.[60–68]

### Diabetes

Diabetes is a prime example of a disease that incorporates intricate multiorgan, hierarchical physiology through vast connections with other highly clustered subnetworks within the human physiome. One example of this is the discovery that in addition to energy metabolism altering bone remodeling, bone hormonally regulates energy metabolism through osteoblast produced osteocalcin (OCN).[8] OCN increases β-cell proliferation, insulin secretion, and insulin sensitivity,[7] and serum OCN levels have been found to be significantly decreased in patients with type 2 diabetes (T2DM).[69] Further studies have shown that specific SNPs in the OCN gene, bone γ-carboxyglutamate protein (BGLAP) gene, are associated with insulin sensitivity and glucose-mediated glucose disposal in African Americans; although no SNP was directly associated with T2DM.[70]

The molecular-level connections between 1,25-dihydroxyvitamin D (1,25-D), OCN, and energy metabolism are multiple including 1,25-D directly increasing the expression of the BGLAP gene[71] and expression of both the vitamin D receptor (VDR) and 1,25-D dependent calcium binding proteins in β-cells and pancreatic tissues, respectively.[72–74] 1,25-D stimulates insulin production possibly through direct modulation of β-cell growth,[75] and 1,25-D supplementation has been shown to improve insulin resistance and glucose tolerance in T2DM.[76–78] Additionally, the immune system is known to play an important role in both the onset and progression of type 1 diabetes and T2DM. Increasing knowledge about the immunomodulating properties of 1,25-D[79–87] in addition to its role in bone remodeling provides a dynamic perspective on the holistic actions of this hormone. These studies raise many questions about the bone-β-cell connection. Are OCN and 1,25-D the important nodes that link the skeletal system, immune system, and energy metabolism? Do other unknown, nonintuitive connections exist that reduce the length between what previously were thought as separate modules within the human body? Would a complex systems analysis show that the edges among OCN, 1,25-D, and glucose control define the physiome as a small-world network? Clearly, a more global understanding of the interplay among the skeletal system, immune system, and energy metabolism is needed.

### Critical illness

With the prioritization of the cardiopulmonary systems in the intensive care unit (ICU), the skeletal system is mostly ignored when dealing with critically ill patients. In addition, a vast array of metabolic derangements characterizes the typical ICU patient including hypothyroidism, adrenal insufficiency, and renal insufficiency, all of which are managed accordingly. However, from a system's perspective, the incorporation of bone as well as other ignored or unknown subsystems in the care of each patient may help improve the morbidity and mortality associated with critical illness.

The skeletal system in chronic critical illness (CCI) is characterized by (1) secondary hyperparathyroidism with predominant vitamin D deficiency ± chronic kidney diseases, (2) PTH suppression with predominant bone hyperresorption and hypercalciuria, or (3) normal PTH levels with a combination of vitamin D deficiency and bone hyperresorption.[88] Systemic hypercytokinemia and a universal uncoupling of the immune-neuroendocrine axis (INA) and various hypothalamic–pituitary axes (HPA) contribute significantly to the bone hyperresorption seen in CCI. Previous studies show as many as 92% of critically ill patients to have bone hyperresorption as shown by increased N-telopeptide levels, a marker for measuring bone resorption.[89] Due to a multitude of factors, additional metabolic bone derangements have been shown in the critically ill population including a deficiency of 25- and 1,25-D in as many as 91% of ICU patients.[89]

So what bone-centric therapies exist that may help slow down the progression of pathological bone resorption with eventual reversal of the catabolic state to one of anabolism? The complex and intricate connections providing the structure to the critical illness model argues that a systems biology based approach may be appropriate to help correct the multiple deficits present. In support of this argument is that individual end-organ hormonal treatments prove to be ineffective or inconclusive at best in combating the hormonal deregulation of the critically ill individual.[90–94] Therefore, therapeutic intervention designed to correct the multiple hormonal deficits at more proximate targets theoretically will lead to downstream correction of a larger subset of actions. The uncoupling of the INA and HPA during prolonged critical illness is one potential interventional target to help curb these gross hormonal derangements. Van den Berghe and colleagues[95] show that co-administration of growth hormone releasing peptide-2 (GHRP-2), thyroid releasing hormone (TRH), and gonadotropin releasing hormone (GnRH) reactivated the abnormal HPA seen in prolonged critically ill male patients. They demonstrate an overall reduction in systemic catabolism and increase in anabolism. These important findings lend support to a systems biology based approach to patient care in the ICU.

Now known to be much more than a calcium and phosphorous regulating hormone, 1,25-D has been found to have multiple immunomodulating effects including suppression of interferon-γ and interleukin-2 activity,[96–98] downregulation of T-lymphocyte proliferation,[96–98] and induction of the antimicrobial peptide cathelicidin.[99] 1,25-D deficiency has been shown to lead to an overall disinhibition of the immune system[79–81] and an indirect decrease in insulin secretion via OCN.[71] Additionally, improved insulin responses occur with administration of 1,25-D.[12] Repletion of 1,25-D in 1,25-D deficient critically ill patients has intriguing potential to help contribute to the re-establishment of normal integrative physiology among the bone, fat, immune system, and β-cells as well as other potential organs effectively facilitating recovery.

### Cardiovascular disease

Using a systems biology approach, one can envision many indirect connections between the CV system and bone. Both the CV system and bone are influenced by, among others, mineralization,[100] estrogen and testosterone,[101–103] 1,25-D,[104–107] vitamin K,[108,109] leptin,[6,110] and catecholamines.[111] Immunomodulating cytokines, such as IL-6, mediate systemic atherosclerosis, leading to CV mortality,[112] as well as osteoclast-stimulated bone hyperresorption.[113] Individually, CV disease and osteoporotic fractures exert a hefty financial toll on the medical system and commonly affect the same population of patients. Dual-role medications that may reduce both the progression of CV disease and osteoporosis may substantially decrease morbidity and mortality while limiting health care expenses.

Recently new research postulates the role of the sympathetic nervous system in bone metabolism, via the adipocyte-derived hormone leptin.[6,114,115] It is now known that leptin, acting through a central hypothalamic relay involving the arcuate nucleus, causes activation of the sympathetic nervous system, which in turn reduces bone formation.[6] Leptin inhibition may therefore have salutary effects on both the CV and skeletal systems. However, this concept remains problematic as peripherally administered leptin has direct stimulatory effects on bone growth[116,117] as well as stimulation of stromal cell differentiation to osteoblasts instead of adipocytes.[118] Clearly, more experimental research is needed to further understanding of the direct and indirect role of leptin on bone metabolism, but this research can be guided by information derived from network analysis and systems biology.

Alternatively, adrenergic receptor blockade can stimulate bone anabolism.[111] In the CV system, β-blockers are routinely used as antihypertensives through their effects on blocking the rennin–angiotensin system as well as peripheral vascular adrenergic receptors. Additionally, β-blockers are a mainstay of treatment for stable angina and acute coronary syndrome as they decrease myocardial oxygen through reducing heart rate, myocardial contractility, and ventricular wall stress and are also commonly used in patients with heart failure and systolic dysfunction to improve symptoms and increase survival. *In vivo* animal studies done by Takeda and colleagues[10] show that mice treated with propanolol after bilateral ovarectomy prevents estrogen deficient bone loss through increased bone formation, an effect later found to be dose-dependent in rat studies.[119]

However, clinical studies have not been as clear. A study using the Framingham osteoporosis cohort shows an increase in bone mineral density in patients taking β-blockers with specific β$_2$-adrenergic receptor polymorphisms, concluding that the effect may be beneficial in specific genotypic populations.[120] Additionally, Bonnet and colleagues[121] found that women treated with β-blockers have increased bone mineral density as well as a decreased rate of fracture if over the age of 40 years with the greatest effects in women over the age of 70 years. Despite this evidence, several other studies present conflicting results,[111,122–125] and the overall effect of

β-blockers on fracture risk, bone mineral density, and prevention of osteoporosis is still unknown.

Another widely used class of CV drugs thought to possibly be beneficial in preventing osteoporosis and fracture risk is the lipid lowering HMG-CoA reductase inhibitors or statins. After the first case-control studies[126,127] showed a reduction in osteoporotic-related fractures in patients taking statins versus other lipid-lowering drugs or placebos, a series of subsequent studies demonstrate conflicting results for fracture risk,[128–134] bone mineral density,[135–139] and bone turnover markers.[140–143]

## Cancer

The landmark study by Garland and Garland[144] showing increased mortality from colon cancer in the northeast United States proposed the idea that vitamin D may reduce the risk of colorectal cancer. Additional epidemiological studies have also shown 1,25-D deficiency to be associated with other types of cancer including breast[145] and prostate.[146] An increase in cancer risk and cancer-related mortality was found with lower levels of serum 25-D in a prospective study-derived predictive model applied to over 47,000 men in the Health Professionals Follow-up Study.[147] This study also shows that men with higher levels of 25-D have a lower risk of developing colorectal, esophageal, pancreatic, and leukemic cancers.[147] However, randomized controlled trials[148–151] and meta-analyses of prospective studies[152–154] looking at the association of 25-D with cancer risk propose mixed results. The highly publicized results from the Women's Health Initiative[148,149] failed to show a decrease in cancer risk; however, the low daily vitamin D supplementation (400 IU/day) used and the significant number of participants reporting a compliance rate as low as 40% combine to call in to question the lack of anticipated anti-cancer benefits.

With the recent findings that 1,25-D has various anti-inflammatory, anti-proliferative, anti-angiogenic, and pro-differentiating effects,[155] much interest has been directed toward 1,25-D for both prevention and treatment of different cancers. Table 3 lists the 1,25-D regulated genes in specific cancers. *In vitro* studies have shown high doses of 1,25-D to be potent inhibitors of tumor progression.[156–158] Calcitriol has been found to have synergistic properties when combined with various platinum agents, taxanes, topoisomerase

**Table 3.** 1,25-D gene regulation in cancer

| Cancer type | Positively regulated | Negatively regulated |
|---|---|---|
| Prostate | p21, p27, IGFBP-3, IGFBP-5, E-cadherin, DUSP-10, TRR1, SOD2, AR, 24-OHase, Metallothioneins | CDK2, MMP-2 |
| Colon | p21, p27, E-cadherin, ZO-1, ZO-2, 24-OHase, CAT1, CAT2, c-Jun, JunB, JunD, FREAC-1/Fox1, ZNF-4/KOX7, Plectin, Filamin, K13, Kallikrein 10, Protease M | c-myc, PPARδ, Tcf-1, CD44, Cyclin D1, Cyclin E, CDK2, CDK6, |
| Breast | p21, p27, TGF-β1, HoxB4, IGFBP-3, IGFBP-5, 24-OHase | CDK2, CDK4, Cyclin D1, Cyclin A, Cyclin D3, c-myc, MMP-9 |
| Leukemia/ myelodysplastic syndrome | PKCα, PKCβI/βIII, PKCγ, PLD1, PLD2, PLCβ, PLCγ, PLA$_2$, COX, LOX, KSR-1, hOC, hOP, 24OHase, p21, p27, Bax, C/EBPβ, TGF-β | p38α/β, BCL-2, |
| Skin Cancer | TGase 1, Involucrin, 24-OHase, IL-10R, PLC-β1, PLC-γ1, PLC-δ1, TGF-β1, Calcium receptor, c-Fos, VDUP1, Protease M, Cystatin M, Amphiregulin, Stromelysin I, Collagenase I | EGF-R, c-myc, K16, SCCA, CRABPII, N-cadherin |

This table shows the genes positively regulated by 1,25-D through the VDR and negatively regulated in specific cancer types (adapted from[173]).

inhibitors, anthracyclines, and antimetabolites via *in vivo* clonogenic assays.[159] Promising results from a phase II trial, ASCENT I (Androgen-Independent Prostate Cancer Study of Calcitriol Enhancing Taxotere),[160] showed a survival advantage in prostate cancer patients taking DN-101 (calcitriol analogue) plus docetaxel versus placebo plus docetaxel. Yet, the promise of vitamin D-analogues as adjunctive prostate cancer treatments was compromised by the discontinuation of the phase III trial, ASCENT II, which compared docetaxel/calcitriol/prednisone to docetaxel/placebo/prednisone due to an increased death rate in the experimental group.[159] Additional phase I trials to determine the toxicity of adjunctive calcitriol with different cancer drugs are underway. However, there do not exist any current well-designed advanced phase trials proving the efficacy of this treatment. Therefore, further investigation is needed.

The prodifferentiating effects of 1,25-D on myeloid cells[161,162] have been exploited in clinical trials looking at 1,25-D treated patients with acute myeloid leukemias (AML) and myelodysplastic syndromes (MDS).[163–167] Despite 1,25-D therapy achieving partial differentiation of blast cells in some of these patients, the clinical results proved

modest at best. Additionally, side effects such as hypercalcemia limit the application of 1,25-D in hematologic malignancies. To combat this, 1,25-D analogues, paricalcitol, and doxercalciferol, were developed to take advantage of the antitumor effects while simultaneously decreasing the hypercalcemic side effects. Unfortunately, clinical trials looking at their effects on AML and MDS have proven less than ideal.[168,169]

## Conclusions

With the development of vast digital databases at the levels of the genome, transcriptome, proteome, metabolome, and physiome, the ability to characterize the biology of the human body at various scales is becoming a reality. This wealth of information helps define explicitly the development of human disease and in turn, facilitates novel approaches for prevention and treatment. Applying this methodology to the skeletal system appears to be an attractive starting point, as it has complex interactions both intraskeletal and as more and more research has shown, extra-skeletal as well. Combining reductionist-derived information with complex systems analysis-derived data of a bone physiome could prove useful in generating novel research queries that could help treat such complex bone-related diseases as osteoporosis and even more systemically defined illnesses such as CV disease and cancer.

## Acknowledgment

We thank Seth Berger for his comments and contributions.

## Conflicts of interest

None of the authors have potential conflicts of interest to disclose.

## References

1. Feero, W.G., A.E. Guttmacher & F.S. Collins. 2010. Genomic medicine—an updated primer. *N. Engl. J. Med.* **362:** 2001–2011.
2. Kreindler, J.L. 2010. Cystic fibrosis: exploiting its genetic basis in the hunt for new therapies. *Pharmacol. Ther.* **125:** 219–229.
3. Kitano, H. 2004. Biological robustness. *Nat. Rev. Genet.* **5:** 826–837.
4. Hillenmeyer, M.E. *et al.* 2008. The chemical genomic portrait of yeast: uncovering a phenotype for all genes. *Science* **320:** 362–365.
5. Kohl, P., E.J. Crampin, T.A. Quinn & D. Noble. 2010. Systems biology: an approach. *Clin. Pharmacol. Ther.* **88:** 25–33.
6. Ducy, P. *et al.* 2000. Leptin inhibits bone formation through a hypothalamic relay: a central control of bone mass. *Cell* **100:** 197–207.
7. Lee, N.K. & G. Karsenty. 2008. Reciprocal regulation of bone and energy metabolism. *Trends Endocrinol. Metab.* **19:** 161–166.
8. Lee, N.K. *et al.* 2007. Endocrine regulation of energy metabolism by the skeleton. *Cell* **130:** 456–469.
9. Elefteriou, F. *et al.* 2005. Leptin regulation of bone resorption by the sympathetic nervous system and CART. *Nature* **434:** 514–520.
10. Takeda, S. *et al.* 2002. Leptin regulates bone formation via the sympathetic nervous system. *Cell* **111:** 305–317.
11. Elefteriou, F. *et al.* 2004. Serum leptin level is a regulator of bone mass. *Proc. Natl. Acad. Sci. USA* **101:** 3258–3263.
12. Karsenty, G. 2006. Convergence between bone and energy homeostases: leptin regulation of bone mass. *Cell. Metab.* **4:** 341–348.
13. Ideker, T., T. Galitski & L. Hood. 2001. A new approach to decoding life: systems biology. *Annu. Rev. Genomics. Hum. Genet.* **2:** 343–372.
14. Bassingthwaighte, J., P. Hunter & D. Noble. 2009. The Cardiac Physiome: perspectives for the future. *Exp. Physiol.* **94:** 597–605.
15. Fenner, J.W. *et al.* 2008. The EuroPhysiome, STEP and a roadmap for the virtual physiological human. *Philos. Transact. A. Math. Phys. Eng. Sci.* **366:** 2979–2999.
16. Hunter, P. *et al.* A vision and strategy for the virtual physiological human in 2010 and beyond. *Philos. Transact. A. Math. Phys. Eng. Sci.* **368:** 2595–2614.
17. Lusis, A.J. & J.N. Weiss. 2010. Cardiovascular networks: systems-based approaches to cardiovascular disease. *Circulation* **121:** 157–170.
18. Newman, M.E. 2001. Scientific collaboration networks. I. Network construction and fundamental results. *Phys. Rev. E. Stat. Nonlin. Soft. Matter Phys.* **64:** 016131-1–016131-8.
19. Stelzl, U. *et al.* 2005. A human protein-protein interaction network: a resource for annotating the proteome. *Cell* **122:** 957–968.
20. Tarassov, K. *et al.* 2008. An in vivo map of the yeast protein interactome. *Science* **320:** 1465–1470.
21. Hopkins, A.L. 2008. Network pharmacology: the next paradigm in drug discovery. *Nat. Chem. Biol.* **4:** 682–690.
22. Keiser, M.J. *et al.* 2007. Predicting new molecular targets for known drugs. *Nature* **462:** 175–181.
23. Garcia, M. & T. McDowell. 2010. Mapping social capital: a critical contextual approach for working with low-status families. *J. Marital Fam. Ther.* **36:** 96–107.
24. Luke, D.A. & J.K. Harris. 2007. Network analysis in public health: history, methods, and applications. *Annu. Rev. Public Health* **28:** 69–93.
25. Salathe, M. & J.H. Jones. Dynamics and control of diseases in networks with community structure. *PLoS. Comput. Biol.* **6:** 1–11.
26. Decker, J.E. *et al.* 2009. Resolving the evolution of extant and extinct ruminants with high-throughput phylogenomics. *Proc. Natl. Acad. Sci. USA* **106:** 18644–18649.
27. Glattfelder, J.B. & S. Battiston. 2009. Backbone of complex networks of corporations: the flow of control. *Phys. Rev. E. Stat. Nonlin. Soft. Matter Phys.* **80:** 036104-1–036104-12.
28. Son, S.W., B.J. Kim, H. Hong & H. Jeong. 2009. Dynamics and directionality in complex networks. *Phys. Rev. Lett.* **103:** 228702-1–228702-4.
29. Yi, H. 2010. Quantum effects on criticality of an Ising model in scale-free networks: Beyond mean-field universality class. *Phys. Rev. E. Stat. Nonlin. Soft. Matter Phys.* **81:** 012103.
30. Kuznetsov, V.A., G.D. Knott & R.F. Bonner. 2002. General statistics of stochastic process of gene expression in eukaryotic cells. *Genetics* **161:** 1321–1332.
31. Ueda, H.R. *et al.* 2004. Universality and flexibility in gene expression from bacteria to human. *Proc. Natl. Acad. Sci. USA* **101:** 3765–3769.
32. Jeong, H., S.P. Mason, A.L. Barabasi & Z.N. Oltvai. 2001. Lethality and centrality in protein networks. *Nature* **411:** 41–42.
33. Berg, J., M. Lassig & A. Wagner. 2004. Structure and evolution of protein interaction networks: a statistical model for link dynamics and gene duplications. *BMC Evol. Biol.* **4:** 51.
34. Giot, L. *et al.* 2003. A protein interaction map of Drosophila melanogaster. *Science* **302:** 1727–1736.
35. Yook, S.H., Z.N. Oltvai & A.L. Barabasi. 2004. Functional and topological characterization of protein interaction networks. *Proteomics* **4:** 928–942.

36. Jeong, H., B. Tombor, R. Albert, *et al.* 2000. The large-scale organization of metabolic networks. *Nature* **407:** 651–654.

37. Loscalzo, J., I. Kohane & A.L. Barabasi. 2007. Human disease classification in the postgenomic era: a complex systems approach to human pathobiology. *Mol. Syst. Biol.* **3:** 124.

38. Barabasi, A.L. & Z.N. Oltvai. 2004. Network biology: understanding the cell's functional organization. *Nat. Rev. Genet.* **5:** 101–113.

39. Barabasi, A.L. & R. Albert. 1999. Emergence of scaling in random networks. *Science* **286:** 509–512.

40. Albert, R. & A. Barabasi. 2002. Statistical mechanics of complex networks. *Rev. Modern Phys.* **74:** 47–97.

41. Cohen, R., K. Erez, ben-Avraham, D. & S. Havlin. 2000. Resilience of the internet to random breakdowns. *Phys. Rev. Lett.* **85:** 4626–4628.

42. Watts, D.J. & S.H. Strogatz. 1998. Collective dynamics of 'small-world' networks. *Nature* **393:** 440–442.

43. Wagner, A. & D.A. Fell. 2001. The small world inside large metabolic networks. *Proc. Biol. Sci.* **268:** 1803–1810.

44. Milo, R. *et al.* 2002. Network motifs: simple building blocks of complex networks. *Science* **298:** 824–827.

45. Basso, K. *et al.* 2005. Reverse engineering of regulatory networks in human B cells. *Nat. Genet.* **37:** 382–390.

46. Gardner, T.S., D. di Bernardo, D. Lorenz & J.J. Collins. 2003. Inferring genetic networks and identifying compound mode of action via expression profiling. *Science* **301:** 102–105.

47. Tyson, J.J., K. Chen & B. Novak. 2001. Network dynamics and cell physiology. *Nat. Rev. Mol. Cell. Biol.* **2:** 908–916.

48. Yu, J., V.A. Smith, P.P. Wang, *et al.* 2004. Advances to Bayesian network inference for generating causal networks from observational biological data. *Bioinformatics* **20:** 3594–3603.

49. Wurthner, J.U., A.K. Mukhopadhyay & C.J. Peimann. 2000. A cellular automaton model of cellular signal transduction. *Comput. Biol. Med.* **30:** 1–21.

50. Post, T.M., S.C. Cremers, T. Kerbusch & M. Danhof. 2010. Bone physiology, disease and treatment: towards disease system analysis in osteoporosis. *Clin. Pharmacokinet.* **49:** 89–118.

51. Dallas, S.L. & L.F. Bonewald. 2010. Dynamics of the transition from osteoblast to osteocyte. *Ann. N. Y. Acad. Sci.* **1192:** 437–443.

52. Dallas, S.L. *et al.* 2009. Time lapse imaging techniques for comparison of mineralization dynamics in primary murine osteoblasts and the late osteoblast/early osteocyte-like cell line MLO-A5. *Cells Tissues Organs* **189:** 6–11.

53. Dean, A.K., S.E. Harris, I. Kalajzic & J. Ruan. 2009. A systems biology approach to the identification and analysis of transcriptional regulatory networks in osteocytes. *BMC Bioinformatics* **10**(Suppl 9): S1–S5.

54. Paic, F. *et al.* 2009. Identification of differentially expressed genes between osteoblasts and osteocytes. *Bone* **45:** 682–692.

55. Gajendran, V.K., J.R. Lin & D.P. Fyhrie. 2007. An application of bioinformatics and text mining to the discovery of novel genes related to bone biology. *Bone* **40:** 1378–1388.

56. Foundation, N.O. 2008. *Physician's Guide to Prevention and Treatment of Osteoporosis.* National Osteoporosis Foundation. Washington, DC.

57. Cummings, S.R. *et al.* 2009. Denosumab for prevention of fractures in postmenopausal women with osteoporosis. *N. Engl. J. Med.* **361:** 756–765.

58. Kendler, D.L. *et al.* 2010. Effects of denosumab on bone mineral density and bone turnover in postmenopausal women transitioning from alendronate therapy. *J. Bone Miner Res.* **25:** 72–81.

59. Reid, I. *et al.* 2010. Effects of denosumab on bone histomorphometry: the freedom and stand studies. *J. Bone Miner Res.* **25:** 2256–2265.

60. Bone, H.G. *et al.* 2010. Odanacatib, a cathepsin-K inhibitor for osteoporosis: a two-year study in postmenopausal women with low bone density. 2010. *J. Bone Miner Res.* **25:** 937–947.

61. Christiansen, C. *et al.* 2010. Safety of bazedoxifene in a randomized, double-blind, placebo- and active-controlled phase 3 study of postmenopausal women with osteoporosis. *BMC Musculoskelet. Disord.* **11:** 130.

62. Grinspoon, S., L. Thomas, K. Miller, *et al.* 2002. Effects of recombinant human IGF-I and oral contraceptive administration on bone density in anorexia nervosa. *J. Clin. Endocrinol. Metab.* **87:** 2883–2891.

63. Hannon, R.A. *et al.* 2010. Effects of the Src kinase inhibitor saracatinib (AZD0530) on bone turnover in healthy men: a randomized, double-blind, placebo-controlled, multiple-ascending-dose phase I trial. *J. Bone Miner Res.* **25:** 463–471.

64. Meriggiola, M.C. *et al.* 2008. Effects of testosterone undecanoate administered alone or in combination with letrozole or dutasteride in female to male transsexuals. *J. Sex Med.* **5:** 2442–2453.

65. Meunier, P.J. *et al.* 2004. The effects of strontium ranelate on the risk of vertebral fracture in women with postmenopausal osteoporosis. *N. Engl. J. Med.* **350:** 459–468.

66. Padhi, D., G. Jang, B. Stouch, *et al.* 2010. Single-dose, placebo-controlled, randomized study of AMG 785: a sclerostin monoclonal antibody. *J. Bone Miner Res.* [Epub ahead of print].

67. Pouwels, S. *et al.* 2010. Use of organic nitrates and the risk of hip fracture: a population-based case-control study. *J. Clin. Endocrinol. Metab.* **95:** 1924–1931.

68. Ruckle, J. *et al.* 2009. Single-dose, randomized, double-blind, placebo-controlled study of ACE-011 (ActRIIA-IgG1) in postmenopausal women. *J. Bone Miner Res.* **24:** 744–752.

69. Rosato, M.T., S.H. Schneider & S.A. Shapses. 1998. Bone turnover and insulin-like growth factor I levels increase after improved glycemic control in noninsulin-dependent diabetes mellitus. *Calcif. Tissue Int.* **63:** 107–111.

70. Das, S.K., N.K. Sharma & S.C. Elbein 2010. Analysis of osteocalcin as a candidate gene for type 2 diabetes (T2D) and intermediate traits in Caucasians and African Americans. *Dis. Markers* **28:** 281–286.

71. Price, P.A. & S.A. Baukol. 1980. 1,25-Dihydroxyvitamin D3 increases synthesis of the vitamin K-dependent bone protein by osteosarcoma cells. *J. Biol. Chem.* **255:** 11660–11663.

72. Ishida, H. & A.W. Norman. 1988. Demonstration of a high affinity receptor for 1,25-dihydroxyvitamin D3 in rat pancreas. *Mol. Cell. Endocrinol.* **60:** 109–117.

73. Johnson, J.A., J.P. Grande, P.C. Roche & R. Kumar. 1994. Immunohistochemical localization of the 1,25(OH)2D3 receptor and calbindin D28k in human and rat pancreas. *Am. J. Physiol.* **267:** E356–E360.

74. Morrissey, R.L., T.J. Bucci, B. Richard, *et al.* 1975. Calcium-binding protein: its cellular localization in jejunum, kidney and pancreas. *Proc. Soc. Exp. Biol. Med.* **149:** 56–60.

75. Lee, S., S.A. Clark, R.K. Gill & S. Christakos. 1994. 1,25-Dihydroxyvitamin D3 and pancreatic beta-cell function: vitamin D receptors, gene expression, and insulin secretion. *Endocrinology* **134:** 1602–1610.

76. Borissova, A.M., T. Tankova, G. Kirilov, *et al.* 2003. The effect of vitamin D3 on insulin secretion and peripheral insulin sensitivity in type 2 diabetic patients. *Int. J. Clin. Pract.* **57:** 258–261.

77. Isaia, G., R. Giorgino & S. Adami. 2001. High prevalence of hypovitaminosis D in female type 2 diabetic population. *Diabetes Care* **24:** 1496.

78. Kumar, S. *et al.* 1994. Improvement in glucose tolerance and beta-cell function in a patient with vitamin D deficiency during treatment with vitamin D. *Postgrad. Med. J.* **70:** 440–443.

79. Chiu, K.C., A. Chu, V.L. Go & M.F. Saad. 2004. Hypovitaminosis D is associated with insulin resistance and beta cell dysfunction. *Am. J. Clin. Nutr.* **79:** 820–825.

80. Giulietti, A. *et al.* 2004. Vitamin D deficiency in early life accelerates Type 1 diabetes in non-obese diabetic mice. *Diabetologia* **47:** 451–462.

81. Hewison, M., M.A. Gacad, J. Lemire & J.S. Adams. 2001. Vitamin D as a cytokine and hematopoetic factor. *Rev. Endocr. Metab. Disord.* **2:** 217–227.

82. Boonstra, A. *et al.* 2001. 1alpha,25-Dihydroxyvitamin d3 has a direct effect on naive CD4(+) T cells to enhance the development of Th2 cells. *J. Immunol.* **167:** 4974–4980.

83. Griffin, M.D. *et al.* 2001. Dendritic cell modulation by 1alpha,25 dihydroxyvitamin D3 and its analogs: a vitamin D receptor-dependent pathway that promotes a persistent state of immaturity in vitro and in vivo. *Proc. Natl. Acad. Sci. USA* **98:** 6800–6805.

84. Lemire, J.M., J.S. Adams, R. Sakai & S.C. Jordan. 1984. 1 alpha,25-dihydroxyvitamin D3 suppresses proliferation and immunoglobulin production by normal human peripheral blood mononuclear cells. *J. Clin. Invest.* **74:** 657–661.

85. Muller, K., C. Heilmann, L.K. Poulsen, *et al.* 1991. The role of monocytes and T cells in 1,25-dihydroxyvitamin D3 mediated inhibition of B cell function in vitro. *Immunopharmacology* **21:** 121–128.

86. Overbergh, L. *et al.* 2000. 1alpha,25-dihydroxyvitamin D3 induces an autoantigen-specific T-helper 1/T-helper 2 immune shift in NOD mice immunized with GAD65 (p524–543). *Diabetes* **49:** 1301–1307.

87. Piemonti, L. *et al.* 2000. Vitamin D3 affects differentiation, maturation, and function of human monocyte-derived dendritic cells. *J. Immunol.* **164:** 4443–4451.

88. Mechanick, J.I. & E.M. Brett. 2002. Endocrine and metabolic issues in the management of the chronically critically ill patient. *Crit. Care. Clin.* **18:** 619–641: viii.

89. Nierman, D.M. & J.I. Mechanick. 1998. Bone hyperresorption is prevalent in chronically critically ill patients. *Chest* **114:** 1122–1128.

90. Takala, J. *et al.* 1999. Increased mortality associated with growth hormone treatment in critically ill adults. *N. Engl. J. Med.* **341:** 785–792.

91. Roberts, I. *et al.* 2004. Effect of intravenous corticosteroids on death within 14 days in 10008 adults with clinically significant head injury (MRC CRASH trial): randomised placebo-controlled trial. *Lancet* **364:** 1321–1328.

92. Minneci, P.C., K.J. Deans, S.M. Banks, *et al.* 2004. Meta-analysis: the effect of steroids on survival and shock during sepsis depends on the dose. *Ann. Intern. Med.* **141:** 47–56.

93. Stathatos, N., C. Levetan, K.D. Burman & L. Wartofsky. 2001. The controversy of the treatment of critically ill patients with thyroid hormone. *Best. Pract. Res. Clin. Endocrinol. Metab.* **15:** 465–478.

94. Angele, M.K., A. Ayala, W.G. Cioffi, *et al.* 1998. Testosterone: the culprit for producing splenocyte immune depression after trauma hemorrhage. *Am. J. Physiol.* **274:** C1530–C1536.

95. Van Den Berghe, G. *et al.* 2002. The combined administration of GH-releasing peptide-2 (GHRP-2), TRH and GnRH to men with prolonged critical illness evokes superior endocrine and metabolic effects compared to treatment with GHRP-2 alone. *Clin. Endocrinol. (Oxf)* **56:** 655–669.

96. Rizzato, G. 1998. Clinical impact of bone and calcium metabolism changes in sarcoidosis. *Thorax* **53:** 425–429.

97. Inui, N. *et al.* 2001. Correlation between 25-hydroxyvitamin D3 1 alpha-hydroxylase gene expression in alveolar macrophages and the activity of sarcoidosis. *Am. J. Med.* **110:** 687–693.

98. Gardner, D.G. 2001. Hypercalcemia and sarcoidosis–another piece of the puzzle falls into place. *Am. J. Med.* **110:** 736–737.

99. Jeng, L. *et al.* 2009. Alterations in vitamin D status and anti-microbial peptide levels in patients in the intensive care unit with sepsis. *J. Transl. Med.* **7:** 1–9.

100. Hofbauer, L.C., C.C. Brueck, C.M. Shanahan, *et al.* 2007. Vascular calcification and osteoporosis–from clinical observation towards molecular understanding. *Osteoporos Int.* **18:** 251–259.

101. Mendelsohn, M.E. & R.H. Karas. 1999. The protective effects of estrogen on the cardiovascular system. *N. Engl. J. Med.* **340:** 1801–1811.

102. Jankowska, E.A. *et al.* 2009. Bone mineral status and bone loss over time in men with chronic systolic heart failure and their clinical and hormonal determinants. *Eur. J. Heart Fail.* **11:** 28–38.

103. Bassil, N., S. Alkaade & J.E. Morley. 2009. The benefits and risks of testosterone replacement therapy: a review. *Ther. Clin. Risk. Manag.* **5:** 427–448.

104. Raggi, P. & M. Kleerekoper. 2008. Contribution of bone and mineral abnormalities to cardiovascular disease in patients with chronic kidney disease. *Clin. J. Am. Soc. Nephrol.* **3:** 836–843.

105. Wang, L., J.E. Manson, Y. Song & H.D. Sesso. 2010. Systematic review: Vitamin D and calcium supplementation in prevention of cardiovascular events. *Ann. Intern. Med.* **152:** 315–323.

106. Ginde, A.A., R. Scragg, R.S. Schwartz & C.A. Camargo, Jr. 2009. Prospective study of serum 25-hydroxyvitamin D level, cardiovascular disease mortality, and all-cause mortality in older U.S. adults. *J. Am. Geriatr. Soc.* **57:** 1595–1603.

107. Melamed, M.L., E.D. Michos, W. Post & B. Astor. 2008. 25-hydroxyvitamin D levels and the risk of mortality in the general population. *Arch. Intern. Med.* **168:** 1629–1637.

108. Shearer, M.J. 2000. Role of vitamin K and Gla proteins in the pathophysiology of osteoporosis and vascular calcification. *Curr. Opin. Clin. Nutr. Metab. Care* **3:** 433–438.

109. Krueger, T., R. Westenfeld, L. Schurgers & V. Brandenburg. 2009. Coagulation meets calcification: the vitamin K system. *Int. J. Artif. Organs* **32:** 67–74.

110. Ahima, R.S. 2004. Body fat, leptin, and hypothalamic amenorrhea. *N. Engl. J. Med.* **351:** 959–962.

111. Graham, S., Hammond-Jones D., Z. Gamie, *et al.* 2008. The effect of beta-blockers on bone metabolism as potential drugs under investigation for osteoporosis and fracture healing. *Expert Opin Investig Drugs* **17:** 1281–1299.

112. Pai, J.K. *et al.* 2004. Inflammatory markers and the risk of coronary heart disease in men and women. *N. Engl. J. Med.* **351:** 2599–2610.

113. Papanicolaou, D.A., R.L. Wilder, S.C. Manolagas & G.P. Chrousos. 1998. The pathophysiologic roles of interleukin-6 in human disease. *Ann. Intern. Med.* **128:** 127–137.

114. Cirmanova, V., M. Bayer, L. Starka & K. Zajickova. 2008. The effect of leptin on bone: an evolving concept of action. *Physiol Res.* **57**(Suppl 1): S143–S151.

115. Elmquist, J.K. & G.J. Strewler. 2005. Physiology: do neural signals remodel bone? *Nature* **434:** 447–448.

116. Steppan, C.M., D.T. Crawford, K.L. Chidsey-Frink, *et al.* 2000. Leptin is a potent stimulator of bone growth in ob/ob mice. *Regul. Pept.* **92:** 73–78.

117. Bagnasco, M., M.G. Dube, A. Katz, *et al.* 2003. Leptin expression in hypothalamic PVN reverses dietary obesity and hyperinsulinemia but stimulates ghrelin. *Obes. Res.* **11:** 1463–1470.

118. Thomas, T. *et al.* 1999. Leptin acts on human marrow stromal cells to enhance differentiation to osteoblasts and to inhibit differentiation to adipocytes. *Endocrinology* **140:** 1630–1638.

119. Bonnet, N. *et al.* 2006. Dose effects of propranolol on cancellous and cortical bone in ovariectomized adult rats. *J. Pharmacol Exp. Ther.* **318:** 1118–1127.

120. Ferrari, S.L. *et al.* 2005. Beta 2 adrenergic receptor, beta-blockers and their influence on bone mass in humans: the Framingham osteoporosis study. *J. Bone Miner Res.* **20:** S11.

121. Bonnet, N. *et al.* 2007. Protective effect of beta blockers in postmenopausal women: influence on fractures, bone density, micro and macroarchitecture. *Bone* **40:** 1209–1216.

122. Rejnmark, L. *et al.* 2004. Fracture risk in perimenopausal women treated with beta-blockers. *Calcif. Tissue Int.* **75:** 365–372.

123. Reid, I.R. *et al.* 2005. beta-Blocker use, BMD, and fractures in the study of osteoporotic fractures. *J. Bone. Miner. Res.* **20:** 613–618.

124. Levasseur, R., P. Dargent-Molina, J.P. Sabatier, *et al.* 2005. Beta-blocker use, bone mineral density, and fracture risk in older women: results from the Epidemiologie de l'Osteoporose prospective study. *J. Am. Geriatr. Soc.* **53:** 550–552.

125. Reid, I.R. *et al.* 2005. Effects of a beta-blocker on bone turnover in normal postmenopausal women: a randomized controlled trial. *J. Clin. Endocrinol Metab.* **90:** 5212–5216.

126. Meier, C.R., R.G. Schlienger, M.E. Kraenzlin, *et al.* 2000. HMG-CoA reductase inhibitors and the risk of fractures. *JAMA* **283:** 3205–3210.

127. Wang, P.S., D.H. Solomon, H. Mogun & J. Avorn. 2000. HMG-CoA reductase inhibitors and the risk of hip fractures in elderly patients. *JAMA* **283:** 3211–3216.

128. Gurm, H.S. & B. Hoogwerf. 2003. The Heart Protection Study: high-risk patients benefit from statins, regardless of LDL-C level. *Cleve Clin. J. Med.* **70:** 991–997.

129. LaCroix, A.Z. *et al.* 2003. Statin use, clinical fracture, and bone density in postmenopausal women: results from the Women's Health Initiative Observational Study. *Ann. Intern. Med.* **139:** 97–104.

130. Reid, I.R., A. Tonkin & C.P. Cannon. 2005. Comparison of the effects of pravastatin and atorvastatin on fracture incidence in the PROVE IT-TIMI 22 trial–secondary analysis of a randomized controlled trial. *Bone* **37:** 190–191.

131. Rejnmark, L. *et al.* 2004. Hip fracture risk in statin users–a population-based Danish case-control study. *Osteoporos Int.* **15:** 452–458.

132. Schoofs, M.W. *et al.* 2004. HMG-CoA reductase inhibitors and the risk of vertebral fracture. *J. Bone. Miner. Res.* **19:** 1525–1530.

133. Scranton, R.E. *et al.* 2005. Statin use and fracture risk: study of a US veterans population. *Arch. Intern. Med.* **165:** 2007–2012.

134. Bauer, D.C. *et al.* 2004. Use of statins and fracture: results of 4 prospective studies and cumulative meta-analysis of observational studies and controlled trials. *Arch. Intern. Med.* **164:** 146–152.

135. Kano, K., K. Nishikura, Y. Yamada & O. Arisaka. 2005. No effect of fluvastatin on the bone mineral density of children with minimal change glomerulonephritis and some focal mesangial cell proliferation, other than an ameliorating effect on their proteinuria. *Clin. Nephrol.* **63:** 74–79.

136. Nakashima, A., R. Nakashima, T. Ito, *et al.* 2004. HMG-CoA reductase inhibitors prevent bone loss in patients with Type 2 diabetes mellitus. *Diabet. Med.* **21:** 1020–1024.

137. Solomon, D.H., J.S. Finkelstein, P.S. Wang & J. Avorn. 2005. Statin lipid-lowering drugs and bone mineral density. *Pharmacoepidemiol. Drug. Saf.* **14:** 219–226.

138. Tanriverdi, H.A., A. Barut & S. Sarikaya. 2005. Statins have additive effects to vertebral bone mineral density in combination with risedronate in hypercholesterolemic postmenopausal women. *Eur. J. Obstet. Gynecol. Reprod. Biol.* **120:** 63–68.

139. Tikiz, C., H. Tikiz, F. Taneli, *et al.* 2005. Effects of simvastatin on bone mineral density and remodeling parameters

in postmenopausal osteopenic subjects: 1-year follow-up study. *Clin. Rheumatol.* **24:** 447–452.

140. Berthold, H.K. *et al.* 2004. Age-dependent effects of atorvastatin on biochemical bone turnover markers: a randomized controlled trial in postmenopausal women. *Osteoporos. Int.* **15:** 459–467.

141. Braatvedt, G.D., W. Bagg, G. Gamble, *et al.* 2004. The effect of atorvastatin on markers of bone turnover in patients with type 2 diabetes. *Bone* **35:** 766–770.

142. Rejnmark, L. *et al.* 2004. Effects of simvastatin on bone turnover and BMD: a 1-year randomized controlled trial in postmenopausal osteopenic women. *J. Bone Miner. Res.* **19:** 737–744.

143. Rosenson, R.S. *et al.* 2005. Short-term reduction in bone markers with high-dose simvastatin. *Osteoporos. Int.* **16:** 1272–1276.

144. Garland, C.F. & F.C. Garland. 1980. Do sunlight and vitamin D reduce the likelihood of colon cancer? *Int. J. Epidemiol.* **9:** 227–231.

145. Bertone-Johnson, E.R. *et al.* 2005. Plasma 25-hydroxyvitamin D and 1,25-dihydroxyvitamin D and risk of breast cancer. *Cancer Epidemiol. Biomarkers Prev.* **14:** 1991–1997.

146. Ahonen, M.H., L. Tenkanen, L. Teppo, *et al.* 2000. Prostate cancer risk and prediagnostic serum 25-hydroxyvitamin D levels (Finland). *Cancer Causes Control* **11:** 847–852.

147. Giovannucci, E. *et al.* 2006. Prospective study of predictors of vitamin D status and cancer incidence and mortality in men. *J. Natl. Cancer Inst.* **98:** 451–459.

148. Chlebowski, R.T. *et al.* 2008. Calcium plus vitamin D supplementation and the risk of breast cancer. *J. Natl. Cancer Inst.* **100:** 1581–1591.

149. Wactawski-Wende, J. *et al.* 2006. Calcium plus vitamin D supplementation and the risk of colorectal cancer. *N. Engl. J. Med.* **354:** 684–696.

150. Lappe, J.M., D. Travers-Gustafson, K.M. Davies, *et al.* 2007. Vitamin D and calcium supplementation reduces cancer risk: results of a randomized trial. *Am. J. Clin. Nutr.* **85:** 1586–1591.

151. Trivedi, D.P., R. Doll & K.T. Khaw. 2003. Effect of four monthly oral vitamin D3 (cholecalciferol) supplementation on fractures and mortality in men and women living in the community: randomised double blind controlled trial. *BMJ* **326:** 1–6.

152. Yin, L. *et al.* Meta-analysis: serum vitamin D and breast cancer risk. *Eur. J. Cancer* **46:** 2196–2205.

153. Yin, L. *et al.* 2009. Meta-analysis: longitudinal studies of serum vitamin D and colorectal cancer risk. *Aliment. Pharmacol. Ther.* **30:** 113–125.

154. Yin, L., E. Raum, U. Haug, *et al.* 2009. Meta-analysis of longitudinal studies: Serum vitamin D and prostate cancer risk. *Cancer Epidemiol.* **33:** 435–445.

155. Deeb, K.K., D.L. Trump & C.S. Johnson. 2007. Vitamin D signalling pathways in cancer: potential for anticancer therapeutics. *Nat. Rev. Cancer* **7:** 684–700.

156. Abe, E. *et al.* 1981. Differentiation of mouse myeloid leukemia cells induced by 1 alpha,25-dihydroxyvitamin D3. *Proc. Natl. Acad. Sci. USA* **78:** 4990–4994.

157. Eisman, J.A., I. Macintyre, T.J. Martin, *et al.* 1980. Normal and malignant breast tissue is a target organ for 1,25-(0H)2 vitamin D3. *Clin. Endocrinol. (Oxf)* **13:** 267–272.

158. Eisman, J.A. *et al.* 1980. 1,25-Dihydroxyvitamin D3 receptor in a cultured human breast cancer cell line (MCF 7 cells). *Biochem. Biophys. Res. Commun.* **93:** 9–15.

159. Trump, D.L., K.K. Deeb & C.S. Johnson. 2010. Vitamin D: considerations in the continued development as an agent for cancer prevention and therapy. *Cancer J* **16:** 1–9.

160. Beer, T.M. *et al.* 2007. Double-blinded randomized study of high-dose calcitriol plus docetaxel compared with placebo plus docetaxel in androgen-independent prostate cancer: a report from the ASCENT Investigators. *J. Clin. Oncol.* **25:** 669–674.

161. Koeffler, H.P. 1983. Induction of differentiation of human acute myelogenous leukemia cells: therapeutic implications. *Blood* **62:** 709–721.

162. Miyaura, C. *et al.* 1981. 1 alpha,25-Dihydroxyvitamin D3 induces differentiation of human myeloid leukemia cells. *Biochem. Biophys. Res. Commun.* **102:** 937–943.

163. Kelsey, S.M. *et al.* 1992. Sustained haematological response to high-dose oral alfacalcidol in patients with myelodysplastic syndromes. *Lancet* **340:** 316–317.

164. Koeffler, H.P., K. Hirji & L. Itri. 1985. 1,25-Dihydroxyvitamin D3: in vivo and in vitro effects on human preleukemic and leukemic cells. *Cancer Treat. Rep.* **69:** 1399–1407.

165. Hellstrom, E. *et al.* 1990. Treatment of myelodysplastic syndromes with retinoic acid and 1 alpha-hydroxy-vitamin D3 in combination with low-dose ara-C is not superior to ara-C alone. Results from a randomized study. The Scandinavian Myelodysplasia Group (SMG). *Eur. J. Haematol.* **45:** 255–261.

166. Slapak, C.A., J.F. Desforges, T. Fogaren & K.B. Miller. 1992. Treatment of acute myeloid leukemia in the elderly with low-dose cytarabine, hydroxyurea, and calcitriol. *Am. J. Hematol.* **41:** 178–183.

167. Ferrero, D. *et al.* 1996. Combined differentiating therapy for myelodysplastic syndromes: a phase II study. *Leuk. Res.* **20:** 867–876.

168. Koeffler, H.P., N. Aslanian & J. O'Kelly. 2005. Vitamin D(2) analog (Paricalcitol; Zemplar) for treatment of myelodysplastic syndrome. *Leuk. Res.* **29:** 1259–1262.

169. Okamoto, R., T. Akagi & P. Koeffler. 2008. Vitamin D compounds and myelodysplastic syndrome. *Leuk. Lymphoma.* **49:** 12–13.

170. Klein, R.J. *et al.* 2005. Complement factor H polymorphism in age-related macular degeneration. *Science* **308:** 385–389.

171. Sladek, R. *et al.* 2007. A genome-wide association study identifies novel risk loci for type 2 diabetes. *Nature* **445:** 881–885.

172. Wellcome Trust Case Control Consortium. 2007. Genome-wide association study of 14,000 cases of seven common diseases and 3,000 shared controls. *Nature* **447:** 661–678.

173. Nagpal, S., S. Na & R. Rathnachalam. 2005. Noncalcemic actions of vitamin D receptor ligands. *Endocr. Rev.* **26:** 662–687.

Ann. N.Y. Acad. Sci. ISSN 0077-8923

# Recent advances in the biology and therapy of muscle wasting

David Glass[1] and Ronenn Roubenoff[2]

[1]Muscle Disease Group and [2]Musculoskeletal Translational Medicine, Novartis Institutes for Biomedical Research, Cambridge, Massachusetts

Address for correspondence: Ronnenn Roubenoff, Musculoskeletal Translational Medicine, Novartis Institutes for Biomedical Research, 220 Massachusetts Avenue, Cambridge, MA 02139. ronenn.roubenoff@novartis.com

The recent advances in our understanding of the biology of muscle, and how anabolic and catabolic stimuli interact to control muscle mass and function, have led to new interest in pharmacological treatment of muscle wasting. Loss of muscle occurs as a consequence of many chronic diseases (cachexia), as well as normal aging (sarcopenia). Although anabolic effects of exercise on muscle have been know for many years, the development of pharmacological treatment for muscle loss is in its infancy. However, there is growing excitement among researchers in this field that developments may yield new treatments for muscle wasting in the future.

Keywords: cachexia; sarcopenia; therapeutics

## Introduction

Loss of muscle is a serious consequence of many chronic diseases and of aging itself, because it leads to weakness, loss of independence, and increased risk of death.[1] Unfortunately, the field suffers from having more definitions than therapies; muscle wasting is an inevitable part of aging, where it is known as sarcopenia (from the Greek, "poverty of flesh").[2] Muscle loss is also common in most organ failure diseases—heart failure, liver or renal failure, chronic obstructive pulmonary disease (COPD)—and in some types of cancer. In these settings, it is known as cachexia (from the Greek, "bad condition").[3] Another term that is commonly used is wasting, such as AIDS wasting, known in Africa as slim disease.[3] Of course, muscle wasting is also inevitable in starvation, where it may be described as kwashiorkor (from the Ga word for second child) or marasmus (from the Greek, "wasting away").

In settings of simple disuse, such as when there is a cast in place or when the patient is subjected to extended bed rest, the term used is simply muscle atrophy, which is also used to describe focal muscle loss, as seen in instances of a variety of compression syndromes or when there is motor neuron injury or disease. Also, there are inflammatory myopathies in which "atrophy" is used to describe the loss of muscle mass. The common denominator to all of these conditions is that the muscle itself is genotypically "normal," in contrast to the dystrophies, in which a genetic mutation, usually but not always in a structural protein, causes an eventual loss of muscle mass due to the degeneration of muscle fibers.

Over the last decade, preclinical studies, mostly on rodent models and cellular studies, have demonstrated a coordinated set of signaling pathways that can modulate muscle mass in the adult mammal.[4] These studies have helped to demonstrate that the various cachexias might have distinct causes, but that they eventually signal into conserved pathways that modulate the breakdown of the sarcomere, perturb protein synthesis, and block the differentiation of a satellite cell or myoblast into a multinucleated fiber.

### Body composition in healthy persons

Under normal conditions, as children enter puberty they develop more muscle mass and strength under the influence of growth hormones, insulin-like growth factor-1 (IGF-1), and sex steroids. In boys, androgen exposure leads to relatively greater muscle growth and less body fat than does estrogen in girls

doi: 10.1111/j.1749-6632.2010.05809.x

in the presence of adequate but not excessive energy intake. An individual's own normative amount of fat and protein stores has somewhat arbitrarily been classified in studies from the 1950s to define the "Standard Man" and "Standard Woman."[5] Standard Man is 25 years old, 154 cm tall, and weighs 70 kg, with about 20% body fat. Standard Woman is also 25 years old, 140 cm tall, and weighs 60 kg, with 25% body fat. Most adults in developed societies continue to gain weight after puberty with about 75% of this weight gain as fat, leading to the well-recognized epidemic of obesity in most parts of the world. Body weight tends to peak between age 55 and 60 years, and then declines slowly over the remaining lifespan. It is not as well recognized, however, that muscle mass generally peaks in the 20s or 30s and then declines. Ironically, weight gain due to obesity offsets this decline to some extent, but only temporarily, and at a substantial health cost.

Loss of more than approximately 40% of body cell mass is fatal.[1] This has been demonstrated in starvation by Jewish physicians in the Warsaw Ghetto during World War II,[6] and in AIDS wasting before the advent of highly active antiretroviral therapy.[7] Smaller losses, however, have demonstrable effects on functional status, immune function, and cancer mortality; for example,[8,9] immune suppression is demonstrable with as little as a 5% loss of lean mass.[10]

Although the treatment of muscle wasting due to starvation is simple in principle—feeding the patient—it may be complicated by metabolic derangements known as the refeeding syndrome.[11] However, for muscle wasting due to cachexia or sarcopenia, the only effective treatment has been a combination of adequate dietary protein, energy, and progressive resistance exercise.[12] This approach, although clearly efficacious in clinical trial settings, has limited effectiveness in the "real world" because of difficulties with expanding exercise to the level of a public health intervention. Thus, there is a great need for new medical treatments that could lead to muscle mass repletion, increase in strength, and enhancement of the effects of diet and exercise in creating muscle anabolism to drive improved function and survival. Recent developments in our understanding of the biology of muscle have opened the possibility that such drugs could be created.

## Clinical syndromes of muscle wasting

Despite the plethora of definitions described earlier, involuntary loss of muscle has recently devolved to two major categories: cachexia and sarcopenia. Cachexia may be defined as a multifactorial syndrome characterized by severe body weight, fat and muscle loss, and increased protein catabolism due to underlying disease(s). Cachexia is clinically relevant because it increases patients' morbidity and mortality.[13] In general, cachexia is a complication of a disease such as heart failure, COPD, or cancer. Inadequate dietary intake may occur, but it is not prominent, and some types of cachexia can occur in the face of actual weight gain (because of accumulation of fat or water). The condition is characterized by the loss of muscle mass and strength and is probably caused partly by the systemic effects of cytokines (tumor necrosis factor-alpha [TNF-alpha], TNF-like weak promoter of apoptosis [TWEAK], IL-6, etc.) produced in the primary lesions and partly by reduced growth factor levels, for example, IGF-1. Additionally, treatments of the underlying disease may produce a secondary suppression of appetite through induction of nausea, reduced salivary flow, and so on.

### Cancer cachexia

Cancer cachexia is a wasting syndrome that arises during the course of most cancers, most often in incurable patients toward the end stages of life. Cancer cachexia has a large impact on prognosis and quality of life and is one of the most common ultimate causes of mortality in cancer patients. There are currently no agents approved for the treatment of cancer cachexia. However, a number of agents are used off-label to treat the condition, for example, cannabinoids, corticosteroids, nutritional supplements, and progestational agents (megestrol acetate). However, significant unmet needs prevail in the cancer cachexia market, meaning most patients do not receive any therapy.

There is no universally agreed definition of cancer cachexia. This makes it difficult to estimate the cancer cachexia population accurately. It is estimated that 55% of all cancer patients experience weight loss during the course of their disease, ranging from over 80% of patients with gastric and pancreatic cancer to about half of lung and colorectal cancers to less than one-third of breast cancer patients. This corresponds to approximately 1.9 million out of the

3.5 million patients diagnosed with cancer in the United States and European Union in 2009. In addition, it is estimated that approximately 1.3 million incurable cancer patients may be potential candidates for palliative treatment of cachexia in 2009.[14]

## Cardiac cachexia

Chronic heart failure (CHF) is a major public health problem in western countries. CHF carries a devastating prognosis that resembles some types of malignant cancer. Its incidence rises steadily from 0.02 per 1000 population per year in those aged 25–34 years to 11.6 per 1000 in those aged 85 years or older. Despite substantial improvements in the management of the disease, the prognosis remains poor, especially in advanced stages of the disease. About half of patients diagnosed with CHF die within four years of diagnosis. The prognosis worsens considerably once cardiac cachexia has been diagnosed. Importantly, the occurrence of cachexia establishes a poor prognosis in patients with CHF, independently of whether the heart failure is regarded as mild or advanced. Mortality at 18 months in unselected patients with CHF in whom cardiac cachexia had been diagnosed was as high as 50%, compared with 17% in non-cachectic patients from the same study population.

The defining clinical characteristic of cachexia is the process of losing weight. This has two consequences: (i) weight loss of a defined degree and possibly over a defined time course should be part of any cachexia definition and (ii) any treatment for cachexia should be able to stabilize or even increase body weight to be considered as "anticachexia therapy." Using data from the studies of left ventricular dysfunction (SOLVD) database, Anker and colleagues[15] suggested a definition for cardiac cachexia as documented nonedematous weight loss of >6% of the previous normal weight observed over a period of >6 months. This definition was validated based on the assumption that a "best" definition should provide the highest sensitivity–specificity product to predict subsequent mortality. In this context, the average weight prior to the onset of heart disease should be used as the previous normal weight.

Unlike the other types of cachexia reviewed here, sarcopenia and cardiac cachexia is characterized primarily by a loss of type I muscle fibers.[16] This teleologically consistent with the idea that due to inadequate oxygen delivery caused by heart failure, the body sacrifices the more oxygen-demanding, highly capillarized, mitochondria-rich type I fibers to reduce oxygen demand. However, when oxygen delivery and demand come into balance with therapy, the cachexia is not corrected automatically. Interestingly, it can be improved with exercise treatment, which has a major clinical benefit in cardiac cachexia without any material effect on cardiac output *per se.*[16]

## COPD cachexia

One of the major problems of patients with COPD is exercise intolerance, which in turn leads to muscle wasting. Muscle wasting should be considered as a serious complication in COPD that has important implications for survival. Disuse, hypoxemia, malnutrition, oxidative stress, and systemic inflammation may all cause muscle atrophy. Particularly when systemic inflammation is elevated, muscle wasting becomes a serious complication in COPD. With worldwide prevalence of COPD increasing, COPD is projected to rank fifth as a burden of disease by 2020. COPD patients have a more sedentary lifestyle than their counterparts. In one study, COPD patients spent 64% of their time sitting or lying down, compared with age-matched healthy individuals who spent more than half of their time walking or standing.[17] Muscle wasting can be seen in all stages of COPD, although its prevalence predictably increases with disease severity, as measured by Global Initiative for Chronic Obstructive Lung Disease (GOLD) stage.[18]

## AIDS wasting

Since the earliest reports of human immunodeficiency virus (HIV) disease, undernutrition has been associated with HIV infection, typically with the late stages of the disease, and may advance to severe wasting and cachexia. Specific micronutrient deficiencies are also recognized to occur with HIV infection, but their actual effect on the clinical course of the disease is hard to assess.[19]

Prior to the introduction of highly active antiretroviral therapy (HAART), wasting was the initial AIDS-defining diagnosis in almost one-third of patients. Following widespread use of HAART, the proportion of patients progressing to a diagnosis of AIDS has diminished, but wasting as defined by loss of body cell mass (BCM) remains a clinical risk.[20] Loss of BCM, a prognostic indicator, has been documented in patients on HAART despite viral load reductions. The mechanism of involuntary weight loss in HIV may be paralleled by a similar phenomenon in other diseases, including some malignancies, in which altered metabolism appears to be an adaptation to the underlying illness. One explanation for this phenomenon is diversion of energy to mount an acute phase anti-inflammatory response. Even in patients with relatively preserved immune function, the average increase in resting energy expenditure is 8–11% relative to individuals without HIV. In patients with AIDS, metabolic rates have increased 20–60% above predicted values. Another contributor to abnormal metabolism in patients with HIV involves disturbances in hormone function. In men, hypogonadism represented by low levels of testosterone inhibits protein synthesis needed to maintain muscle mass. In both men and women, reduced hepatic production of insulin-like growth factor-I, a messenger of growth hormone, and growth hormone resistance are implicated in dysregulation of energy expenditure.[21]

### Rheumatoid cachexia

Rheumatoid arthritis (RA) is the most common autoimmune disease in adults, affecting about 1% of the population in developed countries.[22] RA causes sustained increases in several catabolic cytokines, including TNF-α, IL-1β, and IL-6, which in turn cause elevation in resting energy expenditure and increase muscle protein breakdown.[23] Although cytokine inhibitors are effective in suppressing the joint pain and swelling of RA, they do not appear to reverse rheumatoid cachexia.[24] This is consistent with animal data showing that both IL-1 and TNF must be inhibited in order to preserve muscle mass in the face of an inflammatory insult caused by complete Freund's adjuvant.[25] In addition, rheumatoid cachexia is unique because in concert with the loss of muscle mass there is an increase in fat mass, in contrast to the increase in extracellular water that is typical of cachexia in organ failure disorders such as heart failure, chronic kidney disease, or liver disease.[23]

### Sarcopenia

Sarcopenia is the age-related loss of skeletal muscle mass and function that is a major contributor to the development of frailty in the elderly. Unlike cachexia, sarcopenia is universal and is not a disease. A working definition is currently based on reduced lean body mass,[26] but there is a concerted effort ongoing to improve this definition by adding a functional component such as reduced walking speed. The condition is characterized by loss of muscle mass, a decrease in the number of neuromuscular junctions (NMJ), and hence the number of motor units.[12] Type II fast fibers are particularly affected, which may be important in increasing the probability of falls seen in the elderly. Sarcopenia is an important cause of frailty, a geriatric syndrome characterized by difficulty in responding to physiological stress. Fried and colleagues[27] defined frailty on the basis of the following:

(1) shrinking: weight loss, unintentional, of 10 pounds or 5% of body weight in prior year;
(2) weakness: grip strength in the lowest 20% at baseline, adjusted for gender and body mass index;
(3) poor endurance and energy: as indicated by self-report of exhaustion;
(4) slowness: the slowest 20% of the population was defined at baseline, based on time to walk 15 feet, adjusting for gender and standing height; and
(5) low physical activity level: a weighted score of kilocalories expended per week was calculated at baseline, based on each participant's report. The lowest quintile of physical activity was identified for each gender.

This definition of frailty can be caused by many medical conditions, including dementia, anemia, hypothyroidism, and so on. However, if these are excluded, sarcopenia is often the only detectable cause. Recently, the European Society of Parenteral and Enteral Nutrition published guidelines for the diagnosis of age-related sarcopenia based on the combined presence of the two following criteria:

(1) low muscle mass, i.e., a percentage of muscle mass ≥ 2 standard deviations below the mean measured in young adults of the same sex and ethnic background. Subjects aged 18–39 years in the 3rd NHANES population might be used as reference; and
(2) low gait speed, for example, a walking speed below 0.8 m/s in the 4-m walking test. However, it can be replaced by one of the well-established functional tests used locally as being part of the comprehensive geriatric assessment.[28]

The prevalence of sarcopenia-related disability in the >60-year-old population is estimated between 5% and 10%. This translates to a total of 7–14 million patients in the United States and European Union. The additional care costs associated with managing sarcopenia-related disability in the United States in 2000 were estimated at U.S. $18.5 billion.[29] There is currently no treatment for frailty other than improving nutritional and exercise status. Frail patients require, in addition to direct health care, support to maintain an independent lifestyle or in extreme cases require nursing home care with its attendant increase in mortality. Falls leading to significant or severe injury (e.g., hip fracture), requiring hospitalization could be reduced by improving muscle mass and function in frail patients. The overall financial burden of care costs could also be reduced if elderly persons could be kept in their own homes.

## Current therapies for muscle wasting

Exercise—and especially resistance exercise—is the only proven treatment for reversing or preventing muscle wasting. However, there are major difficulties with applying exercise to large populations. First, skilled trainers, adequate facilities, and motivated patients are needed. These are expensive and often unavailable. Second, many patients are too sick to participate in exercise programs effectively, are not used to exercise before they became ill, and do not like to do exercise. Third, exercise must be continued on an ongoing basis to retain effectiveness—it is one thing to start a patient exercising with a trainer in a rehabilitation setting, and quite another to develop long-term exercise habits. Thus, there is a real need for pharmacological treatment if it can be developed.

The track record of pharmacological interventions for muscle wasting is limited, and much of it is disappointing. Table 1 shows the available drugs that have largely been tested only in AIDS wasting.

## Biological control of muscle homeostasis

Experiments involving self-injection of extracts from dog and guinea pig testicles by Charles-Edouard Brown-Sequard in the 19th century demonstrated that a factor in the testes is important for maintaining lean body mass.[30] Testosterone was later purified from bovine testicles, leading to the Nobel Prize in Chemistry in 1939 (No-

belprize.org. July 12, 2010; http://nobelprize.org/ nobel_prizes/chemistry/laureates/1939/). However, the potential effects of testosterone on skeletal muscle remained in dispute, until it was demonstrated in the 1990s that testosterone treatment resulted in a significant increase in skeletal muscle mass in humans.[31] Since then, testosterone has been shown to positively modulate the effects of many types of anabolic and aerobic exercise.[32]

Testosterone exerts its effects mostly by binding the androgen receptor (AR), which results in a conformational change, allowing for association with the receptor's transcriptional coactivator, beta-catenin. The complex of testosterone/AR/beta-catenin translocates into the nucleus and binds to AR-binding DNA sequences, located in the promoter of particular genes, usually causing an increase in transcription.[33] Testosterone has thus been shown to positively regulate the insulin growth factor-1 (IGF-1),[34] Wnt,[35] and a TGF-β family member called myostatin.[36] Noncanonically, meaning independent of the AR/beta-catenin complex, simple binding of testosterone to the AR itself has been shown to modulate the direct activation of a kinase called Akt, which is a component of the signaling pathway induced by IGF-1 that modulates most of its pro-anabolic effects. The testosterone/AR complex also interacts with glucocorticoid receptors, thus inhibiting the catabolic effects of glucocorticoid.[37,38] The ultimate result of testosterone's activity on skeletal muscle is a stimulation of protein synthesis and a decrease in protein degradation, leading to an increase in muscle fiber cross-sectional area and mass.[39]

Anabolic exercise of skeletal muscle results in hypertrophy or an increase in the size as opposed to the number of individual muscle fibers, due at least in part, to the increased local expression of IGF-1 in the exercised muscle.[40,41] Overexpression of IGF-1, using a muscle-specific promoter, is sufficient to increase muscle fiber cross sectional area and mass in transgenic mice.[41,42] Also, addition of IGF-1 protein *in vitro* to differentiated myotubes promotes their hypertrophy.[43] These data indicate that hypertrophy can be mediated by pathways activated by autocrine or paracrine sources of IGF-1. The binding of IGF-1 to its receptor triggers the activation of several intracellular signaling pathways in skeletal muscle, resulting in the activation of the serine/threonine kinase Akt.[44] During adaptive hypertrophy in adult

**Table 1.** Available anabolic treatments for muscle wasting

| Treatment | Comments |
|---|---|
| Anabolic/Androgenic Steroids | |
| Testosterone supplementation | Men witness significant increases in weight and LBM as a result of testosterone replacement. Although women undergoing this type of treatment experience weight gain as well, it is primarily in the form of fat. |
| Nandrolone decanoate | An injectable derivative of 19-nortestosterone, which is approved only for the treatment of anemia associated with chronic renal failure. |
| Oxandrolone | An oral testosterone derivative, oxandrolone is approved by the FDA as a short-term treatment for weight loss incurred in conjunction with surgery, chronic infection, trauma, or prolonged use of corticosteroids. |
| Oxymetholone | An oral agent that is currently approved as a treatment for anemia, but not for wasting. |
| Appetite stimulants | |
| Megesterol acetate | A synthetic progestational agent that is U.S. FDA approved for HIV-associated anorexia. Hypogonadism, adrenal suppression, deep venous thrombosis, hyperglycemia, and avascular necrosis have been reported in patients receiving megestrol acetate. |
| Dronabinol | A synthetic form of delta-9 tetrahydrocannabinol and is approved by the FDA for treatment of HIV-associated anorexia. The most frequently reported adverse effects of dronabinol are euphoria, dizziness, and thinking abnormalities. |
| Cyproheptadine | An increase in energy intake and weight was witnessed only in a small open-label study in patients with HIV-associated weight loss. |
| Protein anabolic agents | |
| Recombinant human growth hormone | Causes significant weight gain, retention of nitrogen and potassium, and increases in lean body mass. Adverse effects include arthralgias, myalgia, puffiness, and diarrhea. |
| Somatropin (Serostim®, injectable) | The only recombinant human growth hormone that is FDA approved for the treatment of AIDS wasting (limited by high cost and its method of administration). Clinical evidence demonstrates increases in LBM and weight, but not in quality of life. |
| Cytokine Modulation | |
| Thalidomide | The most prevalent adverse effects of thalidomide have been somnolence, peripheral neuropathy, hypersensitivity, and neutropenia. |

LBM = lean body mass.

muscle and in IGF1-induced myotube hypertrophy, Akt is phosphorylated and activated.[45] Expression of an activated form of Akt is sufficient to induce skeletal muscle hypertrophy, in large part by increasing protein synthesis through the mammalian target of rapamycin (mTOR). If rapamycin is given to a rat, to block mTOR, muscle hypertrophy can be inhibited[45] in anabolic settings where significant hypertrophy would otherwise occur. Thus, induction of protein synthesis seems to be a key mechanism for inducing muscle fiber hypertrophy.[44] mTOR and thus muscle mass can be affected by other signaling pathways besides IGF-1. For example, activation of the AMP kinase (AMPK) inhibits mTOR, thus causing muscle atrophy.[46] Also, loss of binding partners for mTOR, such as a protein called raptor, which is required for mTOR's activation of the p70S6 kinase protein synthesis pathway results in muscle atrophy.[47]

In acute settings of muscle atrophy and cachexia, there is a transcriptional upregulation of genes encoding, particularly muscle enzymes that mediate the breakdown of sarcomeric proteins and proteins required for muscle maintenance. These enzymes, muscle-specific ring finger protein 1 (MuRF1) and muscle-associated F-box protein (MAFbx, also called Atrogin-1), are in a family of genes called "E3 ubiquitin ligases" because they encode proteins that bind particular substrates and thereby induce ubiquitin binding and degradation of those protein substrates via the proteasome. In rodent models, MuRF1 and MAFbx have been shown to be induced in multiple models of skeletal muscle atrophy and cachexia; furthermore, these genes have begun to be characterized in settings of active muscle atrophy in humans.[44] Under atrophy conditions, MuRF1 causes the selective breakdown of myosin heavy chain (MyHC)[48] and other components of the thick filament of the sarcomere.[49] MAFbx has been demonstrated to induce the ubiquitination of a protein synthesis modulator, eIF3f, which is involved in the initiation of protein translation.[50] Mice that are genetically null for either MuRF1 or MAFbx demonstrate normal muscle under regular conditions, but less muscle atrophy when the muscle is subjected to atrophy stimuli such as denervation.[45]

MuRF1 was shown to be induced by the pro-inflammatory NF-κB pathway.[51] Mice that overexpress the inhibitor of IκB kinase (IKK) in skeletal muscle, and thus have constitutive muscle-specific activation of NF-κB, demonstrate a significant loss of muscle mass and an upregulation of MuRF1. The IKK/NF-κB signaling pathway can be activated by a variety of cytokines that have been shown to be cachectic, including TNF-α,[52,53] IL-1,[52] IL-6,[54] and TWEAK.[55] If MuRF1 null mice are mated to the IKK mice, the amount of muscle loss is significantly reduced.[51] MAFbx can be upregulated by p38, which is itself activated by inflammatory cytokines.[56] MuRF1 and MAFbx can also be modulated by a transcription factor family alternatively called the "forkheads," or "FOXO." FOXO3 expression is sufficient to cause significant muscle atrophy in transgenic animals.[57] In contrast, FOXO1 has been shown to be required for MuRF1 and MAFbx expression, but its overexpression is not sufficient to induce atrophy.[58] These forkhead family members are controlled by Akt. Akt phosphorylation of the FOXO proteins blocks their transport to the nucleus. In contrast, when Akt is deactivated, such as in acute settings of muscle atrophy, the FOXO proteins can translocate to the nucleus, allowing

for MuRF1 and MAFbx transcription. Thus, IGF-1/Akt signaling does more than inducing protein synthesis. It also blocks protein turnover by inhibiting the FOXO proteins and thus the transcriptional upregulation of the E3 ubiquitin ligases. As for the cytokines that induce MuRF1 and MAFbx upregulation, blockade of TWEAK is sufficient to inhibit denervation-induced atrophy.[59] In settings of denervation, the TWEAK cytokine receptor is upregulated on particular muscle fibers, resulting in their atrophy.[60] The finding that removal of the TWEAK receptor, FN14, is sufficient to modulate denervation atrophy suggests the surprising possibility that denervation-induced upregulation of the TWEAK receptor is required for the full induction of the NF-κB pathway, and the atrophy that results from that induction.

PGC-1α is a transcriptional coactivator that has been shown to be sufficient to induce fiber-type switching toward the "slow," more mitochondrial-rich, transcriptional program. Indeed, PGC-1α overexpression directly causes mitochondriogenesis. Slow fibers, such as soleus muscle in rodents, are relatively more resistant to muscle atrophy than fast fibers. Therefore, PGC-1α overexpressing mice were found to be resistant to muscle atrophy.[61] Interestingly, old PGC-1α transgenic mice have recently been shown to have much less of the perturbations that characterize sarcopenic muscle. In contrast to wild-type animals, PGC1-α transgenics maintain muscle mass and function well into old age.[62]

In addition to IGF-1, other secreted proteins have been demonstrated to perturb skeletal muscle size. Myostatin, also called growth and differentiation factor-8 (GDF-8), is a TGF-β family member which is a negative regulator of muscle mass.[63] Myostatin's effect was demonstrated in studies with mice that were made null for the myostatin gene,[64] and also by correlating increases in muscle mass that were observed in strains of cattle with a loss of myostatin;[65–67] the loss of myostatin resulted in more than doubling in muscle mass. It has been suggested that other TGF-β superfamily molecules, distinct from myostatin, play a role in modulating skeletal muscle size because myostatin$^{-/-}$ mice that are mated with mice that are transgenic for follistatin (TG$^{follistatin}$), which is capable of inhibiting not only myostatin but also its close relative GDF-11 and other TGF-β molecules such as the activins, resulted in an even greater increase in muscle size.[68]

*In vitro* studies with myostatin have been performed on rodent cells. In these studies, it has been shown that myostatin can block the differentiation of myoblasts into myotubes.[69–72] Experiments both *in vitro* and *in vivo* have demonstrated that myostatin signals first binds the type II activin receptor, IIb, which then allows for interaction with type I receptors ALK4 or ALK5.[73] The binding of myostatin to these receptor complexes results in the phosphorylation and activation of the transcription factors Smad2 and Smad3, which translocate to the nucleus upon phosphorylation.[74] In a study of myostatin and other TGF-β molecules on human skeletal myoblasts (HuSkMC) and myotubes, HuSkMCs respond to myostatin at physiologic concentrations, 0.1–300 ng/ml, resulting in a decrease in fusion index, myotube diameter, creatine kinase (CK) activity, and expression of MyoD and myogenin.[75] It was previously demonstrated that follistatin, a more general inhibitor of TGF-β molecules, could induce an additive increase in muscle mass when combined with myostatin.[68] A range of other TGF-β molecules are shown to be able to block muscle differentiation, including the more distantly-related activins, bone morphogenic molecule-2 (BMP-2), and TGF-β itself.[75] Myostatin inhibits activation of Akt in both myoblasts and myotubes.[75,76] It was recently reported that muscle-specific ablation of TORC1 (by ablating RAPTOR) results in a dystrophic phenotype.[47] Inhibition of RAPTOR, and thus TORC1 does not by itself block muscle differentiation, but does contribute to myostatin's inhibitory effects by resulting in an increase in myostatin-induced Smad phosphorylation, establishing a feed-forward mechanism: myostatin activates Smad2, which inhibits Akt inhibiting TORC1, which in turn potentiates myostatin's activation of Smad2.[75,76] Addition of IGF-1 dominantly blocks the effects of myostatin when applied to either myoblasts or myotubes.[75] The precise intersection between the two pathways may be multifold, but it is clear that Akt is a particular nexus and that IGF-1 can rescue the activation of the PI3K/Akt pathway that is blunted by myostatin.

## Novel ways to increase muscle mass and function

### Androgens
The demonstration of the mediators of muscle mass suggests novel ways to treat the loss of muscle seen in settings of cachexia, atrophy, and sarcopenia.

During aging, testosterone levels progressively diminish resulting in a significant incidence of hypogonadism.[77] Given the promyotropic effects of testosterone, this age-related loss likely plays a role in the coincident decrease in muscle mass and function that characterizes sarcopenia. However, clinicians must be cautious about giving testosterone itself, because of its pleiotropic effects on multiple tissues. Oxandrolone, a synthetic testosterone analog, has been approved for treatment of nonspecific weight loss and anemia for over 40 years, but it is also nonselective in terms of anabolic versus androgenic effects. Therefore, more selective androgen receptor modulators (SARMs) are being tested in settings of cachexia (see Table 1). These SARMs have been screened for the preservation of positive effects on skeletal muscle versus other tissues where androgen stimulation is less desired such as the prostrate gland. At the moment, several new compounds are in development and results are expected over the next 2–3 years, including Ostarine (GTX-024, GTX Pharma, Nashville, TN, USA), MK-0773 (Merck, Whitehouse Station, PA, USA), BMS-564929 (Bristol-Myers Squibb, Princeton, NJ, USA), LGD-4033 (Ligand Pharmaceuticals), GSK971086 (GlaxoSmithKline, London, England), and others.

### Growth hormone
Another pathway that has been explored primarily in the treatment of AIDS wasting and fat redistribution is growth hormone treatment. The effect has been shown to be statistically significant, and growth hormone (GH) is approved by the U.S. FDA for treatment of AIDS wasting (Serostim, Merck Serono). However, the effect is not very large, and treatment of the underlying disease with highly effective antiretroviral agents has largely eliminated AIDS wasting as a clinical problem in the developed countries. Ironically, in the developing world where wasting remains a problem, GH is too expensive to be prescribed.

An indirect way of increasing GH secretion is via GH secretagogues. The prototype, ghrelin, is a 28-amino acid peptide that is produced by cells in the stomach, intestines, and hypothalamus. It is also produced in lesser amounts by the placenta, kidney, and pituitary gland. When released by neuroendocrine cells of the hypothalamus, it gains access to the anterior pituitary via the hypothalamo–hypophyseal portal tract, binds to receptors on somatotropes and causes the release of GH into the circulation. In addition to its GH-releasing action, ghrelin increases hunger through an action on hypothalamic feeding centers and stimulates gastric emptying, which facilitates increased food intake. Because of their combined anabolic action on skeletal muscle and appetite, stimulant properties ghrelin and low molecular weight agonists of the ghrelin receptor are considered attractive candidates for the treatment of cachexia and sarcopenia.

### Myostatin pathway

More recently, inhibition of myostatin has been attempted by several means. Direct inhibition by an antibody that binds to myostatin, MYO-029 (Wyeth) failed to substantially improve muscle strength or mass in patients with muscular dystrophies.[78] This study has been criticized for recruiting patients with three different types of muscular dystrophy (MD): Becker, FSHD, and limb-girdle MD, with no more than 12 patients in each disease × dose cohort. However, the limitation may in fact be the potency of the molecule. Thus, an alternative approach taken by Acceleron Pharmaceuticals and Novartis is to inhibit the activin receptors, which blocks signal transduction by myostatin as well as several related ligands. These molecules are now in early phase clinical trials.

### Angiotensin and beta adrenoreceptor blockade

Infusion of angiotensin II into rodents leads to muscle wasting, and elevated circulating levels of angiotensin II occur in patients with cachexia secondary to congestive heart failure, chronic renal disease, and cancer. Angiotensin receptor blockade prevents skeletal muscle atrophy in rats with experimentally induced CHF and the chronic use of ACE inhibitors, which block the generation of angiotensin II, results in larger low- extremity muscle mass in elderly patients. ACE inhibitors have also been demonstrated to reduce cachexia in patients with cancer and CHF. The mechanism by which angiotensin II causes skeletal muscle atrophy is incompletely understood but is probably indirect because there are few or no angiotensin II receptors expressed on skeletal muscle cells. In the liver, angiotensin II increases the production of inflammatory mediators such as IL-6 (see above) and serum amyloid A which act synergistically to downregulate IGF-1 signaling in skeletal muscle, thus promoting

proteolysis. In addition, it increases the expression of Atrogin-1, MuRF-1, and two ubiquitin E3 ligases previously reported to be involved in the degradation of structural proteins in various forms of skeletal muscle atrophy. Angiotensin II infusion also reduces the expression of IGF-l in skeletal muscle and liver which further enhances catabolic processes.

Beta-2 adrenoreceptor agonists such as clenbuterol have been abused for many years by athletes performing sports that require muscle power. The ability of beta-2 agonists to cause muscle hypertrophy both *in vivo* and *in vitro* and to block atrophy is well documented as is their ability to cause a concomitant increase in muscle strength in healthy individuals. However, there are relatively few reports of their efficacy in treating patients with muscle weakness probably because of concerns regarding their potential to produce cardiovascular side effects such as tachycardia and arrhythmias. The exact mechanism by which beta-2 agonists induce skeletal muscle hypertrophy is unclear regarding the coupling mechanism and second messenger systems involved, and may differ from the classical $G_s\alpha$ activation that leads to cyclic adenosine monophosphate (cAMP) generation in smooth muscle. More likely, it involves signaling through another G protein such as $G\beta$ or $G\gamma$ with activation of phosphatidylinositol-3-kinase (PI3K) and subsequently AKT leading to protein synthesis through the p70S6 kinase pathway.

### Anticytokine strategies

Finally, anticatabolic strategies, primarily based on trying to inhibit the catabolic cytokines that have been implicated in cachexia, are also being investigated. Anti-TNF therapy, however, does not seem to reverse rheumatoid cachexia despite strong efficacy in suppressing the joint swelling and pain of rheumatoid arthritis.[24] This is consistent with animal data suggesting that in inflammatory cachexia at least, both IL-1 and TNF must be blocked to successfully prevent muscle atrophy. More promising are recent results of IL-6 blockade (ALD518, Alder Pharmaceuticals, Bothell, WA, USA), which showed benefit in nine patients with cachexia and fatigue associated with advanced non-small cell lung cancer.

## Conclusions

The advances in our understanding of muscle biology that have occurred over the past decade have led to new hopes for pharmacological treatment of muscle wasting. These treatments will be tested in humans in the coming years and offer the possibility of treating cachexia and even sarcopenia/frailty. These treatments should be developed in the setting of appropriate dietary and exercise strategies, much as diabetes or atherosclerosis pharmacotherapy is added to a foundation of proper lifestyle interventions. We believe that the synergies between pharmacological, dietary, and exercise treatments will open the way for marked improvements in the outcomes of many chronic diseases and possibly of aging itself.

## Conflicts of interest

The authors are both employees of Novartis.

## References

1. Roubenoff, R. & J. Kehayias. 1991. The meaning and measurement of lean body mass. *Nutr. Rev.* **46:** 163–175.
2. Rosenberg, I. 1989. Summary comments. *Am. J. Clin. Nutr.* **50:** 1231–1233.
3. Roubenoff, R., S. Heymsfield, J. Kehayias, *et al.* 1997. Standardization of nomenclature of body composition in weight loss. *Am. J. Clin. Nutr.* **66:** 192–196.
4. Glass, D.J. 2010. Signaling pathways perturbing muscle mass. *Curr. Opin. Clin. Nutr. Metab. Care* **13:** 225–220.
5. Snyder, W.S., C.M., E.S. Nasset, L.R. Karhausen, *et al.* 1975. Report of the task group on reference man. In *International Commission on Radiation Protection*. Series 23. 1-64. Pergamon Press. Oxford.
6. Winick, M. 1979. *Hunger Disease—Studies by Jewish Physicians in the Warsaw Ghetto.* John Wiley & Sons, NY.
7. Kotler, D., A. Tierney & R. Pierson. 1989. Magnitude of body cell mass depletion and the timing of death from wasting in AIDS. *Am. J. Clin. Nutr.* **50:** 444–447.
8. Dewys W.D. *et al.* 1980. Prognostic effect of weight loss prior to chemotherapy in cancer patients. *Am. J. Med.* **69:** 491–497.
9. Baracos, V. 2006. Cancer-associated cachexia and underlying biological mechanisms. *Ann. Rev. Nutr.* **26:** 435–461.
10. Keusch, G. 2003. The history of nutrition: malnutrition, infection, and immunity. *J. Nutr.* **133:** 336S-340S.
11. Keys, A., J. Brozek, A. Henschel, *et al.* 1950. *The Biology of Human Starvation.* University of Minnesota Press. Minneapolis, MN.
12. Hughes, V. & R. Roubenoff. 2000. Sarcopenia: current concepts. *J. Gerontol. Med. Sci.* **55A:** M716–M724.
13. Evans, W. *et al.* 2008. Cachexia: a new definition. *Clin. Nutr.* **27:** 793–799.
14. Laviano, A., M. Meguid, A. Inui, *et al.* 2005. Therapy insight: cancer anorexia-cachexia syndrome — when all you can eat is yourself. *Nat. Clin. Pract. Oncol.* **2:** 158–165.
15. Anker, S. *et al.* 1997. Wasting as an independent risk factor for mortality in chornic heart failure. *Lancet* **349:** 1050–1053.
16. Pu, C. *et al.* 2001. Randomized trial of progressive resistance training to counteract the myopathy of chronic heart failure. *J. Appl. Physiol.* **90:** 2341–2350.
17. Pitta, F. *et al.* 2005. Characteristics of physical activities in daily life in chronic obstructive pulmonary disease. *Am. J. Resp. Crit. Care. Med.* **171:** 972–977.
18. Scholls, A., R. Broekhuizen, C. Weling-Scheepers, *et al.* 2005. Body composition and mortality in chronic obstructive pulmonary disease. *Am. J. Clin. Nutr.* **82:** 53–59.
19. Tang, A., N. Graham & A. Saah. 1996. Effects of micronutrient intake on survival in human immunodeficiency virus type 1 infection. *Am. J. Epidemiol.* **143:** 1244–1256.
20. Wanke, C. *et al.* 2000. Weight loss and wasting remain common complications in individuals infected with human immunodeficiency virus in the era of highly active antiretroviral therapy. *Clin. Infect. Dis.* **31:** 803–805.
21. Grinspoon, S. *et al.* 1997. Body composition and endocrine function in women with acquired immunodeficiency syndrome. *J. Clin. Endocrinol. Metab.* **82:** 1332–1337.
22. Lawrence, R. *et al.* 2008. Estimates of the prevalence of arthritis and other rheumatic conditions in the United States: part II. *Arthritis Rheum.* **59:** 26–35.
23. Roubenoff, R. *et al.* 1994. Rheumatoid cachexia: cytokine-driven hypermetabolism accompanying reduced body cell mass in chronic inflammation. *J. Clin. Invest.* **93:** 2379–2386.
24. Metsios, G. *et al.* 2007. Blockade of tumor necrosis factor-alpha in rheumatoid arthritis: effects on components of rheumatoid cachexia. *Rheumatol.* **46:** 1824–1827.
25. Hamada, K., E. Vannier, D. Smith, *et al.* 2000. Inflammatory cachexia induces sarcoactive gene expression in a rat model of adjuvant arthritis. *FASEB J.* **14:** A572.
26. Baumgartner, R. *et al.* 1998. Epidemiology of sarcopenia among the elderly in New Mexico. *Am. J. Epidemiol.* **147:** 755–763.
27. Fried, L. *et al.* 2001. Frailty in older adults: evidence for a phenotype. *J. Gerontol.* **56A:** M146–M156.
28. Cruz-Jentoft, A. *et al.* 2010. Sarcopenia: European consensus on definition and diagnosis. *Age and Ageing* **39:** 412–423.
29. Janssen, I., D. Shepard, P. Katzmarzyk & R. Roubenoff. 2004. The health care costs of sarcopenia in the United States. *J. Am. Geriatr. Soc.* **52:** 80–85.
30. Brown-Sequard, C. 1899. Note on the effects produced on man by subcutaneous injections of a liquid obtained from the testicles of animals. *Lancet* **2:** 105–107.
31. Bhasin, S. *et al.* 1996. The effects of supraphysiologic doses of testosterone on muscle size and strength in normal men. *N. Engl. J. Med.* **335:** 1–7.
32. Sattler, F.R. *et al.* 2009. Testosterone and growth hormone improve body composition and muscle performance in older men. *J. Clin. Endocrinol. Metab.* **94:** 1991–2001.
33. Singh, R. *et al.* 2006. Testosterone inhibits adipogenic differentiation in 3T3-L1 cells: nuclear translocation of androgen receptor complex with {beta}-catenin and T-cell factor 4 may bypass canonical Wnt signaling to downregulate adipogenic transcription factors. *Endocrinol.* **147:** 141–154.
34. Ferrando, A.A. *et al.* 2002. Testosterone administration to older men improves muscle function: molecular and physiological mechanisms. *Am. J. Physiol. Endocrinol. Metab.* **282:** E601–E607.

35. Singh, R. *et al.* 2009. Regulation of myogenic differentiation by androgens: cross talk between androgen receptor/{beta}-catenin and follistatin/transforming growth factor-{beta} signaling pathways. *Endocrinol.* **150:** 1259–1268.

36. Mendler, L., Z. Baka, A. Kov·cs-Simon, *et al.* 2007. Androgens negatively regulate myostatin expression in an androgen-dependent skeletal muscle. *Biochem. Biophys. Res. Commun.* **361:** 237–242.

37. Eason, J., S. Dodd & S. Power. 2003. Use of anabolic steroids to attenuate the effects of glucocorticoids on the rat diaphragm. *Phys. Ther.* **83:** 29–36.

38. Zhao, W. *et al.* 2008. Testosterone protects against dexamethasone-induced muscle atrophy, protein degradation and MAFbx upregulation. *J. Steroid. Biochem. Mol. Biol.* **110:** 125–129.

39. Brodsky, I., P. Balagopal & K. Nair. 1996. Effects of testosterone replacement on muscle mass and muscle protein synthesis in hypogonadal men—a clinical research center study. *J. Clin. Endocrinol. Metab.* **81:** 3469–3475.

40. DeVol D.L., P. Rotwin, J.L. Sadow & L.S. Dux. 1990. Activation of insulin-like growth factor gene expression during work-induced skeletal muscle growth. *Am. J. Physiol. Endocrinol. Metab.* **259:** E89–E95.

41. Coleman, M.E. *et al.* 1995. Myogenic vector expression of insulin-like growth factor I stimulates muscle cell differentiation and myofiber hypertrophy in transgenic mice. *J. Biol. Chem.* **270:** 12109–12116.

42. Musaro, A. *et al.* 2001. Localized Igf-1 transgene expression sustains hypertrophy and regeneration in senescent skeletal muscle. *Nat. Genet.* **27:** 195–200.

43. Rommel, C. *et al.* 2001. Mediation of IGF-1-induced skeletal myotube hypertrophy by PI(3)K/Akt/mTOR and PI(3)K/Akt/GSK3 pathways. *Nat. Cell Biol.* **3:** 1009–1013.

44. Glass, D.J. 2005. Skeletal muscle hypertrophy and atrophy signaling pathways. *Int. J. Biochem. Cell Biol.* **37:** 1974–1984.

45. Bodine, S.C. *et al.* 2001. Akt/mTOR pathway is a crucial regulator of skeletal muscle hypertrophy and can prevent muscle atrophy in vivo. *Nat. Cell Biol.* **3:** 1014–1019.

46. Krawiec, B., G. Nystrom, R. Frost, *et al.* 2007. AMP-activated protein kinaes agonists increase mRNA content of the muscle-specific ubiquitin ligases MAFbx and MuRF1 in C2C12 cells. *Am. J. Physiol. Endocrinol. Metab.* **292:** E1555–E1567.

47. Bentzinger, C. *et al.* 2008. Skeletal muscle-specific ablation of raptor, but not of rictor, causes metabolic changes and results in muscle dystrophy. *Cell Metab.* **8:** 411–424.

48. Clarke, B.A. *et al.* 2007. The E3 Ligase MuRF1 degrades myosin heavy chain protein in dexamethasone-treated skeletal muscle. *Cell Metab.* **6:** 376–385.

49. Cohen, S. *et al.* 2009. During muscle atrophy, thick, but not thin, filament components are degraded by MuRF1-dependent ubiquitylation. *J. Cell Biol.* **185:** 1083–1095.

50. Lagirand-Cantaloube, J. *et al.* 2008. The initiation factor eIF3-f is a major target for Atrogin1/MAFbx function in skeletal muscle atrophy. *EMBO J.* **27:** 1266–1276.

51. Cai, D. *et al.* 2004. IKKbeta/NF-kappaB activation causes severe muscle wasting in mice. *Cell* **119:** 285–298.

52. Ninomiya-Tsuji, J. *et al.* 1999. The kinase TAK1 can activate the NIK-I kappaB as well as the MAP kinase cascade in the IL-1 signalling pathway. *Nature* **398:** 252–256.

53. Ladner, K., M. Caligiuri, & D. Guttridge. 2003.Tumor Necrosis Factor-regulated biphasic activation of NF-kappa B is required for cytokine-induced loss of skeletal muscle gene products. *J. Biol. Chem.* **278:** 2294–2303.

54. Jackson-Bernitsas, D. *et al.* 2007. Evidence that TNF-TNFR1-TRADD-TRAF2-RIP-TAK1-IKK pathway mediates constitutive NF-B activation and proliferation in human head and neck squamous cell carcinoma. *Oncogene* **26:** 1385–1397.

55. Dogra, C., H. Changotra, S. Mohan & A. Kumar. 2006. Tumor necrosis factor-like weak inducer of apoptosis inhibits skeletal myogenesis through sustained activation of nuclear factor-kappaB and degradation of MyoD protein. *J. Biol. Chem.* **281:** 10327–10336.

56. Li, Y.-P. *et al.* 2005. TNF-{alpha} acts via p38 MAPK to stimulate expression of the ubiquitin ligase atrogin1/MAFbx in skeletal muscle. *FASEB J* **19:** 362–370.

57. Zhao, J. *et al.* 2007. FoxO3 coordinately activates protein degradation by the autophagic/lysosomal and proteasomal pathways in atrophying muscle cells. *Cell Metab.* **6:** 472–483.

58. Sandri, M. *et al.* 2004. Foxo transcription factors induce the atrophy-related ubiquitin ligase atrogin-1 and cause skeletal muscle atrophy. *Cell* **117:** 399–412.

59. Kumar, M., D. Makonchuk, H. Li, *et al.* 2009. TNF-like weak inducer of apoptosis (TWEAK) activates proinflammatory signaling pathways and gene expression through the activation of TGF-{beta}-activated kinase 1. *J. Immunol.* **182:** 2439–2448.

60. Mittal, A. *et al.* 2010. The TWEAKÀiFn14 system is a critical regulator of denervation-induced skeletal muscle atrophy in mice. *J. Biol. Chem.* **188:** 833–849.

61. Sandri, M. *et al.* 2006. PGC-1{alpha} protects skeletal muscle from atrophy by suppressing FoxO3 action and atrophy-specific gene transcription. *Proc. Natl. Acad. Sci. USA* **103:** 16260–16265.

62. Wenz, T., S. Rossi, L. Rotundo, *et al.* 2009. Increased muscle PGC-1alpha expression protects from sarcopenia and metabolic disease during aging. *Proc. Natl. Acad. Sci. USA* **106:** 20405–20410.

63. Lee, S. *et al.* 2004. Regulation of muscle protein degradation: coordinated control of apoptotic and ubiquitin-proteasome systems by phosphatidylinositol 3 kinase. *J. Am. Soc. Nephrol.* **15:** 1537–1545.

64. McPherron, A. & S. Lee. 1997. Double muscling in cattle due to mutations in the myostatin gene. *Proc. Natl. Acad. Sci. USA* **94:** 12457–12461.

65. Grobet, L. *et al.* 1997. A deletion in the bovine myostatin gene causes the double-muscled phenotype in cattle. *Nat. Genet.* **17:** 71–74.

66. McPherron, A., A. Lawler & S. Lee. 1997. Regulation of skeletal muscle mass in mice by a new TGF-beta superfamily member. *Nature* **387:** 83–90.

67. Kambadur, R., M. Sharma, T. Smith, *et al.* 1997. Mutations in myostatin (GDF8) in double-muscled Belgian blue and Piedmontese cattle. *Genome Res.* **7:** 910–916.

68. Lee, S. 2007. Quadrupling muscle mass in mice by targeting TGFbeta signaling pathways. *PLoS One* **2:** E789.

69. Langley, B. *et al.* 2002. Myostatin inhibits myoblast differentiation by downregulating MyoD expression. *J. Biol. Chem.* **277:** 49831–49840.

70. Rios, R., S. Fernandez-Nocelos, I. Carneiro, *et al.* 2004. Differential response to exogenous and endogenous myostatin in myoblasts suggests that myostatin acts as an autocrine factor *in vivo. Endocrinol.* **145:** 2795–2803.

71. McFarlane, C. *et al.* 2006. Myostatin induces cachexia by activating the ubiquitin proteolytic system through an NF-?B-independent, FoxO1-dependent mechanism. *J. Cell Physiol.* **209:** 501–514.

72. Yang, W., Y. Zhang, Y.-P. Li, *et al.* 2007. Induces cyclin D1 degradation to cause cell cycle arrest through a phosphatidylinositol 3-kinase/AKT/GSK-3beta pathway and is antagonized by insulin-like growth factor 1. *J. Biol. Chem.* **282:** 3799–3808.

73. Tsuchida, K., M. Nakatani, A. Uezumi, *et al.* Signal transduction pathway through activin receptors as a therapeutic target of musculoskeletal diseases and cancer. *Endocrine J.* **55:** 11–21.

74. Rebbapragada, A., H. Benchabane, J. Wrana, *et al.* 2003. Myostatin signals through a transforming growth factor {beta}-like signaling pathway to block adipogenesis. *Mol. Cell Biol.* **23:** 7230–7242.

75. Trendelenburg, A. *et al.* 2009. Myostatin reduces AKT/TORC1/p70S6K signaling, inhibiting myoblast differentiation and myotube size. *Am. J. Physiol. Cell Physiol.* **296:** C1258–C1270.

76. Sartori, R. *et al.* 2009. SMAD2 and SMAD3 transcription factors control muscle mass in adulthood. *Am. J. Physiol. Cell Physiol.* **296:** C1248–C1257.

77. Morley, J. *et al.* 1997. Longitudinal changes in testosterone, leutinizing hormone, and follicle-stimulating hormone in healthy older men. *Metabolism* **46:** 410–413.

78. Wagner, K. *et al.* 2008. A phase I/II trial of MYO-029 in adult subjects with muscular dystrophy. *Ann. Neurol.* **63:** 561–571.

Ann. N.Y. Acad. Sci. ISSN 0077-8923

ANNALS OF THE NEW YORK ACADEMY OF SCIENCES
Issue: *Molecular and Integrative Physiology of the Musculoskeletal System*

# Mechanical loading, cartilage degradation, and arthritis

Hui B. Sun

Leni and Peter W. May Department of Orthopedics, Mount Sinai School of Medicine, New York, New York

Address for correspondence: Hui B. Sun, One Gustave L. Levy Place, Box 1188, Leni and Peter W. May Department of Orthopedics, Mount Sinai School of Medicine, New York, NY 10029. Herb.Sun@mssm.edu

Joint tissues are exquisitely sensitive to their mechanical environment, and mechanical loading may be the most important external factor regulating the development and long-term maintenance of joint tissues. Moderate mechanical loading maintains the integrity of articular cartilage; however, both disuse and overuse can result in cartilage degradation. The irreversible destruction of cartilage is the hallmark of osteoarthritis and rheumatoid arthritis. In these instances of cartilage breakdown, inflammatory cytokines such as interleukin-1 beta and tumor necrosis factor-alpha stimulate the production of matrix metalloproteinases (MMPs) and aggrecanases (ADAMTSs), enzymes that can degrade components of the cartilage extracellular matrix. In order to prevent cartilage destruction, tremendous effort has been expended to design inhibitors of MMP/ADAMTS activity and/or synthesis. To date, however, no effective clinical inhibitors exist. Accumulating evidence suggests that physiologic joint loading helps maintain cartilage integrity; however, the mechanisms by which these mechanical stimuli regulate joint homeostasis are still being elucidated. Identifying mechanosensitive chondroprotective pathways may reveal novel targets or therapeutic strategies in preventing cartilage destruction in joint disease.

Keywords: mechanical loading; cartilage degradation; arthritis

## Introduction

Mechanical force has long been appreciated as a regulator of musculoskeletal tissues, and may be the most important single environmental factor responsible for joint homeostasis. The transmission of mechanical loads requires the participation of multiple joint components, including bone, muscles, articular cartilage, and ligaments/tendons. It has become apparent that these and other joint tissues (e.g., synovium) are sensitive to the magnitude, duration, and nature of mechanical stimuli. In this review, we will first examine the role of nonphysiological joint loading (overuse and disuse) and arthritis in cartilage destruction, and then contrast these findings to the protective nature of physiological loading. Lastly, the review will discuss the underlying mechanisms mediating these effects of loading at the genetic and epigenetic levels and explore the potential use of moderate loading as a chondroprotective therapy.

## Nonphysiological joint loading and cartilage destruction

### Joint overuse and disuse

Although joints maintain homeostasis within a physiological range of mechanical loading, both reduced loading and overloading have catabolic effects, particularly for the cartilaginous components. Injuries of the knee articular cartilage surfaces are frequently observed in athletes and recreational participants of high-impact sports, including football, basketball, and soccer.[1–3] These cartilage injuries frequently result in association with other acute injuries, such as ligament or meniscal injuries, traumatic patellar dislocations, and osteochondral injuries.[4–6] For example, articular cartilage defects are reported in up to 50% of athletes undergoing anterior cruciate ligament reconstruction.[4,7] Articular cartilage damage can also develop from chronic pathological joint-loading patterns, such as joint instability or misalignment.[4–6] These

doi: 10.1111/j.1749-6632.2010.05808.x

articular cartilage injuries, as a consequence of either acute or chronic high-intensity loads, will frequently result in cartilage degeneration, which may eventually lead to osteoarthritis (OA).[8,9]

Studies show that excessive mechanical stress can directly damage the cartilage extracellular matrix (ECM) and shift the balance in chondrocytes to favor catabolic activity over anabolism. While direct measurements of *in vivo* cartilage-on-cartilage contact stresses due to overuse in human joints have not been made, experimental evidence indicates the range of nonphysiological loading intensities. There appears to be a critical threshold of 15–20 megapascals (MPa) for cell death and collagen damage due to a single impact load in bovine cartilage explants.[10] In another study on bovine cartilage explants, chondrocyte apoptosis occurred at peak stresses as low as 4.5 MPa and increased with peak stress in a dose-dependent manner, while degradation of the collagen matrix occurred in the 7–12 MPa range,[11] suggesting chondrocyte apoptosis may precede cartilage matrix damage. In addition to peak stress, strain rate appears to be an important parameter implicated in cartilage damage. Bovine cartilage explants compressed at a strain rate of 0.01/sec to a final strain of 50% showed in no measurable effect on biosynthetic activity or mechanical functionality (compressive and shear stiffness) in chondrocytes, although peak stresses reached 12 MPa,[12] a stress high enough to cause injurious effects such as cell death.[11] However, compression at higher strain rates (0.1 and 1 per second), resulting in peak stresses of about 18 and 23 MPa, reduced total protein biosynthesis and compressive and shear stiffness,[12] suggesting that an increase in peak stress and strain rate is associated with increased injury. Consistent with these results, high strain rates were reported to result in significant matrix fluid pressurization and impact-like surface cracking with cell death near the superficial zone in bovine osteochondral explants.[13] Studies also found that repetitive impact loading of 5 MPa at 0.3 Hz induces collagen network damage and chondrocyte necrosis and apoptosis,[14,15] suggesting that impact damage is cumulative. Taken together, the data indicate that high levels of peak stress, high strain rates, and long-term injurious mechanical loading are usually correlated with deleterious effects on cartilage. However, caution must be used when extrapolating these *in vitro* results to the *in vivo* situation. Indenters used in these experiments are likely to cause large stress amplitudes and gradients at the location of impact, while in normally congruent articular joints, these stresses may be limited due to a more even distribution of force.[15] Therefore, the range of nonphysiological load intensities *in vivo* should be greater than those reported in these *in vitro* studies.

Similarly to overuse, reduced loading, occurring commonly as a result of joint immobilization following spinal cord injuries or secondary to treatments for acute musculoskeletal injury[16,17] or as a result of joint diseases such as arthritis,[18] also creates catabolic responses within articular cartilage. Prolonged immobilization *in vivo* causes cartilage thinning,[19,20] tissue softening,[20,21] and reduced proteoglycan content,[19,22] leading to cartilage matrix fibrillation, ulceration, and erosion.[23,24] In patients with spinal cord injury, articular cartilage atrophies at a rate greater than that reported in age-associated OA.[25] Even during short-term reduced loading conditions such as seven weeks of partial weight bearing at the knee, there exhibited a significant degree of cartilage thinning in all compartments of the knee, although no cartilage lesions were observed.[26]

Cartilage damage due to joint overuse and prolonged disuse is irreversible due to the limited spontaneous repair of articular cartilage.[27–29] The avascular nature of articular cartilage prevents the physiological inflammatory response to tissue injury and resultant repair. This often results in biochemical and metabolic changes resembling early osteoarthritis, such as an accumulation of degradative enzymes and cytokines, disruption of collagen ultrastructure, increased hydration, and fissuring of the articular surface.[30]

## Cartilage destruction in arthritis

OA is a progressive joint disease that affects the structural and functional integrity of articular cartilage, as well as the adjacent bone and other joint tissues. Clinical symptoms associated with OA include joint pain, stiffness, and swelling, which may lead to muscle weakness and impaired physical function.[31] OA is generally diagnosed radiographically by bony changes, including joint space narrowing, development of osteophytes, subchondral sclerosis, and subchondral cyst formation.[32] These clinical and radiographic features are most commonly diagnosed in the knee, hip, and hand joints.[31] Although the cause of OA is unknown, risk factors for developing arthirits include old age, joint trauma, obesity, and heritable genetic factors.[33,34]

Although the most evident morphological sign of OA is the progressive degeneration of articular cartilage, and OA is classified as noninflammatory arthritis, synovitis may also play a role in the progression of cartilage degradation.[35] Joint swelling, mononuclear cell infiltration, and synovial and subchondral angiogenesis are frequently present in OA joints, resulting in[35–39] synovial hypertrophy and hyperplasia. Increased numbers of immune cells (e.g. activated B cells and T lymphocytes) in synovial tissue are also observed,[35,40,41] leading to the induction of cytokines interleukin-1β (IL-1β) and tumor necrosis factor-α (TNF-α),[39] and the production of proteases which target the cartilage extracellular matrix.[40] These cartilage breakdown products can further provoke the release of degradative enzymes from synovial cells.[42] Therefore, in developing therapeutic strategies against cartilage degradation, the effect of the adjoining joint tissues should be considered.

Rheumatoid arthritis (RA) is a systemic inflammatory disease that produces a progressive degeneration of the musculoskeletal system. RA is characterized by both joint inflammation and destruction of bone and articular cartilage.[43] One of the most prevalent chronic conditions, RA is found in approximately 1% of the adult population in the United States.[44] Even with appropriate drug therapy, up to 7% of patients remain disabled to some extent 5 years after disease onset and 50% are too disabled to work 10 years after onset.[44] Consequently, RA results in considerable direct and indirect costs. Disabilities associated with RA lead to a restriction of regular daily activity and diminished quality of life.[45]

The cause of RA is unknown and the pathogenic process may be initiated by several different events. The initial stages of RA involve the initiation and establishment of autoimmunity, followed by an inflammatory response, angiogenesis to maintain the chronic inflammatory state, and tissue degradation of the joint. During the inflammatory process the usually thin layer of synovial lining cells becomes greatly enlarged from the influx of monocytes and lymphocytes from the circulation and from the local proliferation of fibroblasts, giving rise to a synovial hypertrophy called a pannus.[46] This pannus formation contains activated B and T cells,

plasma cells, mast cells, and particularly activated macrophages which are involved in feeding back to promote perpetuation of inflammation. Central to this inflammatory response in the rheumatoid joint are high levels of inflammatory mediators, cytokines, and growth factors,[47] including TNF-α, IL-1β, IL-17, and transforming growth factor (TGF)-β.[48]

New blood vessels form following the inflammatory response, in a process known as angiogenesis, to maintain the inflammatory state by transporting nutrients to the developing pannus and inflammatory cells to sites of synovitis.[49,50] Angiogenesis is regulated by many inducers and inhibitors, including acidic and basic fibroblast growth factors (bFGF), TGF, angiopoietin, placenta growth factor, and vascular endothelial growth factor (VEGF). VEGF, a dimeric glycoprotein which induces formation of new blood vessels and increases vascular permeability,[51] is highly expressed in the synovial tissues of RA patients, and serum levels of VEGF correlate with RA disease activity.[52] VEGF knockout mice show reduced RA disease activity and synovial angiogenesis in antigen-induced models of arthritis,[53] suggesting that inhibiting VEGF-mediated angiogenesis is likely to suppress rheumatoid inflammation.

Although the etiologies of both OA and RA differ, a common consequence of both is cartilage destruction (Fig. 1). This degradation can be attributed to an imbalance between the anabolic and catabolic activities of the chondrocyte, suggesting that alterations in cellular metabolism contribute to the onset and progression of the disease.[54] Catabolism of the cartilage ECM, classically defined by the degradation of both collagen fibrils and proteoglycans,[55] involves a variety of degradative enzymes, many of which are matrix metalloproteinases (MMPs) and ADAMTS (a disintegrin and metalloproteinase with thrombospondin motifs).[56,57] The human genome has 24 MMP genes, including two duplicated MMP-23 genes. The MMP family consists of the collagenases, (MMPs 1, 8, and 13) which degrade collagens types I, II, and III, the gelatinases (MMPs 2 and 9), which target denatured collagen, the stromelysins (MMPs 3, 7, 10, and 11), which degrade several ECM proteins and are involved in proenzyme post-translational activation, the membrane-type MMPs (MT-MMP 1–4), and a diverse subgroup including MMPs 12, 20, and 23.[58]

**Osteoarthritis**                    **Rheumatoid Arthritis**

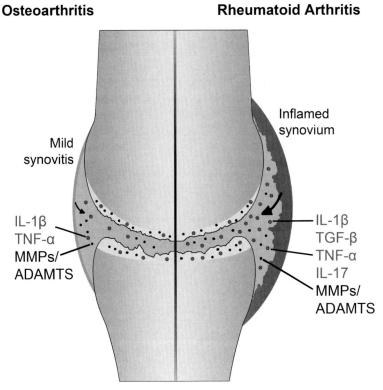

**Figure 1.** Cartilage destruction in osteoarthritis and rheumatoid arthritis. Although the sites of origin for OA and RA differ, cartilage degradation is a common consequence for both. In OA, cartilage breakdown initially occurs at the articular surface and the synovium may produce proteases and cytokines that accelerate the progression of the disease. In RA, the main cellular sources of degradative enzymes are synovial lining cells and macrophages. See text for abbreviations.

While several ADAMTS family members have aggrecanolytic activity (e.g. ADAMTS-4),[59] ADAMTS-5 is considered to be the most active in this regard.[60] However, not all ADAMTSs have been characterized as proteinases, including procollagen N-propeptide processing enzymes (ADAMTS-2, ADAMTS-3, and ADAMTS-14), and a von Willebrand factor cleaving enzyme (ADAMTS-13),[61] and ADAM17, or TNF-α converting enzyme (TACE),[62–65] which releases TNF-α from the macrophage surface.

Active MMPs are bound by tissue inhibitors of metalloproteinases (TIMPs) in a 1:1 ratio to limit degradation of the cartilage matrix.[56,66] The human genome has 4 TIMPs (TIMPs-1 to -4), of which TIMP-3 has the broadest inhibition spectrum, affecting a large number of metalloproteinases, including several ADAMTSs.[66–69,72] If TIMP levels exceed those of active enzyme, connective tissue turnover is prevented. However, a deficiency of TIMPs, as demonstrated in TIMP-3 knockout mice,[73] leads to an imbalance between MMPs and TIMPs, and results in cartilage degradation.

Suppressing elevated levels of MMPs and ADAMTS in arthritis should be regarded to have substantial clinical benefit; however, a safe and effective inhibitor has not yet been developed. MMP inhibitors used in clinical trials have so far all failed to show significant efficacy in human diseases and in some instances these inhibitors resulted in adverse effects such as musculoskeletal pain and joint swelling.[56] The side effects of MMP inhibitors were attributed largely to their lack of selectivity because metalloproteinases share structural similarities.[74] Unintended MMP inhibition has been proven to be problematic, because aside from tissue breakdown, MMPs play critical roles in development, wound healing, and angiogenesis.[75,76] For example, MMP-2 knockout mice exhibited more severe clinical and histological arthritis than wild-type mice in an Ab-induced arthritis model[77] and MMP-3 knockout mice developed OA more readily than the wild-type mice in a joint destabilization model.[78] Deletion of the MMP14[79] or the ADAM15[80] gene in the mouse caused arthritis-like symptoms. An alternative noninvasive approach in suppressing MMP and ADAMTS activity may include moderate mechanical joint stimulation. Reports have demonstrated the efficacy of physiologic joint loading such as passive motion therapy in arthritis.[81–83] Furthermore, observations that moderate loading can antagonize MMP upregulation induced by inflammatory cytokines,[81,82] indicates that such a strategy might be therapeutically useful.

## Physiological joint loading and cartilage homeostasis

### Moderate loading of healthy joints

While reduced loading and overloading both cause cartilage degradation, moderate levels of activity maintain normal cartilage integrity. Physiologic mechanical stimulation of the articular cartilage generates biochemical signals which increase the anabolic activity of the chondrocytes.[84–86]

Numerous in vitro studies have increases in both proteoglycan synthesis and cell proliferation were observed after dynamic stimulation of chondrocytes in cartilage explants, 3D and monolayer cultures.[85–88] Cyclic pressure-induced strain increased expressions of aggrecan and superficial zone protein in monolayer cultures.[89,90] Aggrecan gene expression was also increased by 5 and 10 MPa intermittent hydrostatic pressure, which also stimulated collagen II gene expression.[84] Hydrostatic pressure in bovine cartilage explants loading increased both proteoglycan production and aggrecan mRNA synthesis.[91,92]

### Chondroprotection by mechanical loading

It has been shown that moderate exercise may protect against cartilage degradation in animals that spontaneously develop OA. Hamsters that ran 6–11 km/day maintained normal cartilage integrity, while their sedentary counterparts developed fibrillation, pitting, and fissuring in the articular cartilage during this 3-month study.[93] Normal cage activity rats with a surgically induced model of OA showed more macroscopic and histologic cartilage degradation when compared to exercised OA rats which ran 30 cm/sec for 30 min for 28 days.[94]

In humans, moderate recreational physical exercise is associated with a decreased risk of severe knee OA, suggesting that exercise has a protective effect against developing cartilage degradation.[95] Furthermore, moderate joint loading may improve joint health and function in OA.[96] Accumulating studies have demonstrated the effectiveness of non-drug treatment modalities, for example, exercise and physical activity, as an adjunct to drug therapy in patients with OA and RA.[97] Consequently, clinical practice guidelines developed to aid health practitioners in treating OA [98,99] recommend exercise therapy to reduce pain and improve function, based largely on expert opinion and the results of large randomized controlled trials evaluating exercise.[100–102]

Clinical studies have shown that in RA, moderate exercise has systemic anti-inflammatory effects by reducing disease activity.[103,104] In vitro studies have provided evidence for a direct mechanosensitive, antiinflammatory effect on joint tissues. Mechanical strain in synoviocytes decreases expression of prostaglandin- $E_2$ ($PGE_2$),[105] an inflammatory mediator, and strain and shear both decreased the expressions of MMP-1 and -13 in RA synoviocytes.[106,107] Furthermore, chondrocytes exposed to moderate levels of intermittent hydrostatic pressure inhibit IL-1β-induced matrix degradation.[108]

In addition to exercise therapy, LIPUS (low-intensity pulsed ultrasound) is a mechanical stimulation therapy clinically used in the orthopedic field to treat fractures by accelerating healing.[109–111] LIPUS promotes synthesis of several matrix components in chondrocytes in vitro, including type II collagen, type X collagen, and aggrecan.[112] LIPUS also increased production of type II collagen in an experimental OA rat model and ameliorated histological cartilage damage when compared to untreated groups.[113] In humans, therapeutic ultrasound alleviates pain and helps restore joint function in OA patients,[114] but whether it serves as a chondroprotective agent remains to be investigated.

## Signaling pathways mediating loading effects on cartilage

The regulation of MMPs and ADAMTS by mechanical loading can be both beneficial—in maintaining tissue homeostasis, and deleterious—in inducing pathological conditions. Nonphysiologic loading induces MMP/ADAMTS-mediated damage of the cartilage matrix[115] while physiological loads maintain modest levels of MMP expression.[116] Clearly, mechanical forces have a strong influence on

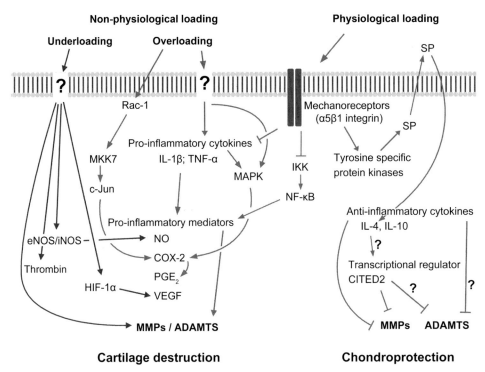

**Figure 2.** Pro- and antiinflammatory mechanisms of mechanical loading on chondrocytes. Nonphysiological loading in the forms of underloading and overloading induce cartilage damage through pathways involving the upregulation of MMPs and ADAMTSs, while physiological loading blocks these increases.

cartilage homeostasis. *In vivo*, articular cartilage is subject to various mechanical forces including dynamic cyclic and static compression, shear stress, tensile strain, and hydrostatic pressure.[117] Likewise, in culture, chondrocytes respond to a range of loading conditions through diverse metabolic responses, and it is from these *in vitro* studies that much of our understanding of the catabolic and anticatabolic pathways has emerged (Fig. 2).

### Proinflammatory role of loading in arthritis

The signal cascades that are activated as a result of non-physiological mechanical stimulation are in many cases similar to those activated by known catabolic stimuli such as proinflammatory cytokines. Mitogen-activated protein kinases (MAPK), activated by IL-1β and TNF-α, are activated by static loading in chondrocytes.[118] High-amplitude dynamic stimulation activates NF-κB, a key element of many chondrocyte signaling pathways, including responses to proinflammatory cytokines[119] and matrix fragments.[120] Mechanical overloading stimulates expression of VEGF, which appears to be necessary for me-

chanically induced MMP-1, -3, and -13 expressions.[121] A VEGF-mediated signaling pathway may also play a role in the upregulation of ADAMTS5 and MMPs 1, 2, 3, 9, and 13 after cartilage injury.[122]

Cyclooxygenase-2 (COX-2) is a pivotal proinflammatory enzyme and aberrant expression of COX-2 protein in articular tissues is an earmark of arthritis.[123–125] COX-2 expression was observed in inflamed, but not normal paw tissue from adjuvant-induced arthritis rats.[125–127] Oral administration of a selective COX-2 inhibitor markedly suppressed COX-2 expression, production of prostanoid PGE$_2$, paw edema, and inflammatory cell infiltration in the joints,[126] suggesting that COX-2 is primarily responsible for the elevated production of prostanoids at sites of disease and inflammation. It was found that COX-2 is induced by high intensities of fluid shear stress (20 dyn/cm$^2$) in chondrocytes in a process involving the Rac/MEKK1/MKK7/JNK2/c-Jun-C/EBPβ pathway.[128]

PGE$_2$ is another major catabolic mediators involved in cartilage degradation and chondrocyte apoptosis.[129–131] Higher levels of PGE2 are observed

in OA cartilage when compared to normal cartilage[132] and mice lacking a PGE2 membrane receptor, EP4, show decreased incidence and severity
of cartilage degradation in collagen-induced arthritis.[133] Following impact with a peak stress of around
25 MPa, levels of PGE2 increased 22-fold in the
medium of explants after 3 days.[134] A subsequent
study found that the MAPK, AP-1, and NF-κB signaling pathways are involved in the upregulation of
PGE$_2$ and proinflammatory mediator NO release by
IL-1β.[135]

While the pathways involved in the destruction
of cartilage by mechanical overloading are being
elucidated, less is known of the mechanisms initiated by underloading. Joint immobilization in
animal models show that cartilage degradation is
mediated by MMPs and ADAMTS.[136–138] These
models also reveal that immobilization increases
expression of HIF-1α, VEGF,[139] prothrombin,[140]
and nitric oxide synthase,[141] but the mechanism

for their upregulation in reduced loading conditions and subsequent effect on cartilage integrity is
unknown.

## Mechanisms of mechanical chondroprotection

Studies demonstrate that biomechanical signals
within a physiologic range of intensity, duration, and frequency are potent antiinflammatory
signals that attenuate proinflammatory gene induction in chondrocytes (Table 1). At low mag-
nitudes *in vitro*, biomechanical signals inhibit IL-
1β- or TNF-α-induced transcriptional activation of
COX-2, MMPs, IL-1β, and other proinflammatory
molecules.[82,119,142–144] *In vivo*, motion-based therapies have been demonstrated to mitigate joint inflammation in animal models of antigen-induced
arthritis. Mechanical signals generated from these
passive joint motion therapies were reported to
suppress expression of proinflammatory catabolic
mediators, such as IL-1β, COX-2, and MMP-1 as

**Table 1.** Protective effects of moderate joint loading on skeletal tissue *in vivo*

| Type of mechanical loading | Animal model/Tissue | Effect | Reference |
|---|---|---|---|
| CPM (24 or 48 h) | Antigen-induced arthritis/Rabbit Menisci | ↓ GAG degradation<br>↓ COX-2<br>↓ MMP-1<br>↓ IL-1β<br>↑ IL-10 | Ferretti *et al.*[82] |
| CPM (24, 48, 96 h, or 12 h/day for 2 weeks) | Antigen-induced arthritis/Rabbit distal femoral cartilage | ↓ Cartilage erosion<br>↓ IL-1β<br>↓ COX-2<br>↓ MMP-1<br>↑ IL-10 | Ferretti *et al.*[81] |
| CPM (1h/day for 1week) | Joint immobilization/Rat distal femoral cartilage | ↓ MMP-3<br>↓ ADAMTS-5<br>↓ GAG degradation | Leong *et al.*[137] |
| Low-intensity pulsed ultrasound (20 min/day for 28 days) | Osteoarthritis/Rat knee articular cartilage | ↑ Collagen type-II synthesis<br>↓ Cartilage degradation (Mankin Score) | Naito *et al.*[113] |
| Treadmill running (30 cm/sec for 30 min for 28 days) | Osteoarthritis/Rat knee articular cartilage | ↓ Cartilage degradation (Mankin Score)<br>↓ Activated Caspase 3 | Galois *et al.*[94] |
| Moderate-intensity exercise | Osteoarthritis/Human knee articular cartilage | ↑ GAG content | Roos *et al.*[96] |

CPM = continuous passive motion; GAG = glycosaminoglycans; COX-2 = cyclooxygenase-2; MMP = matrix metalloproteinase; IL = interleukin.

well as inducing expression of the antiinflammatory cytokine, IL-10.[81,82]

Shear flow regulated MMP-1 and -13 suppress expression in a stress-dependent manner in human chondrocytes. Chondrocytes subject to fluid flow at 5 dyn/cm$^2$ resulted in decreases of MMP-1 and -13 expression and an increase in transcriptional regulator CITED2 (CBP/p300-interacting transactivator with ED-rich tail 2).[145] In contrast, basal levels of CITED2 expression were found at 0 and 20 dyn/cm$^2$, which correlated with increased levels of MMP-1 and -13. Transfecting antisense CITED2 plasmids into chondrocytes abolished the loading-mediated downregulation of MMP-1, suggesting that CITED2 is required by load-driven MMP downregulation. CITED2 overexpression prevented increases of MMP-1 and -13 in the presence of IL-1β.[145] Since CITED2 does not bind DNA directly, but instead regulates target genes by interacting with other transcription factors such as p300, co-immunoprecipitation with co-factor p300 specific antibody was used to identify potential regulatory proteins in p300 binding protein complexes. In cells exposed to moderate flow shear (5 dyn/cm$^2$), p300 was complexed with CITED2 while p300 complexed with MMP transactivator Ets-1[146] in cells under 20 dyn/cm$^2$.[145] This study suggests that CITED2 may mediate the mechanical loading-induced downregulation of MMP by competing with MMP transactivator Ets-1 for limiting amounts of cofactor p300 protein.

Human chondrocyte monolayer cultures subjected to pressure-induced strain significantly inhibited MMP-3 mRNA expression after 1 hour.[89] It was demonstrated that stretch-activated calcium channels, α5β1 integrin, and IL-4 were involved in the mechano-signaling pathway regulating MMP-3.[147,148] Thus strain-induced inhibition of MMP-3 is transduced via an integrin-dependent IL-4 autocrine/paracrine route.[89]

NF-κB transcription factors are involved in both acute and chronic inflammatory responses by regulating a wide range of pro-inflammatory and anti-apoptotic genes. NF-κB is a rapid response, inducible, transcription factor which when activated by upstream pro-inflammatory pathways, translocates to the nucleus,[149] where it induces transcription of its target genes, including pro-inflammatory cytokines and mediators.[150–152] Physiological mechanical stimuli, in the form of cyclic tensile strain,

block IL-1β-induced transcriptional activity of NF-κB by inhibiting multiple steps in the NF-κB signaling cascade.[119,153,154] These findings suggest that mechanical signals use specific target sites to trigger NF-κB signaling.

### Epigenetic regulation of cartilage health

Control of gene expression depends not only on genetic factors (e.g., DNA sequences within promoter regions that recognize specific transcriptional regulators), but also on epigenetic mechanisms, which may play a critical role in cartilage health. Epigenetic regulation includes the activation of normally silent genes through DNA hypomethylation, which allows for an open chromatin structure,[155–157] or silencing of normally expressed genes through DNA hypermethylation, which prevents access of transcription factors to their promoter.[155,156] It has been previously demonstrated that there is hypomethylation at specific sites in the promoters of MMP-3, MMP-9, MMP-13, and ADAMTS-4 in degradative OA chondrocytes.[158,159] On the other hand, hypermethylation of the osteogenic protein 1 (OP-1) promoter in aged chondrocytes is correlated with suppression OP-1.[161] Additionally, loss of DNA methylation is associated with an upregulation of leptin in OA.[160]

In addition to DNA methylation, gene transcription can also be regulated by deacetylation through histone deacetylases (HDACs).[162] HDAC inhibitors (HDACi) have been demonstrated to block the IL-1β induction of key MMPs (e.g., MMP-1, -3, -8 and -13) and aggrecanases (e.g., ADAMTS4, ADAMTS5, and ADAMTS9) at the mRNA level in explant-conditioned culture and consequently inhibit cartilage degradation.[163] In an adjuvant arthritis model, HDAC inhibitors (phenylbutyrate and trichostatin A) induced expression of the cell cycle inhibitors p21[Cip1] and p16[INK4] in synovial cells and inhibited the expression of TNF-α, resulting in reduced joint swelling, a decrease in subintimal mononuclear cell infiltration, inhibition of synovial hyperplasia, suppression of pannus formation, and prevention of cartilage and bone destruction.[164] Therefore, HDAC inhibitors may have therapeutic potential for RA patients by suppressing inflammation and preserving joint function.

Although mechanical loading induced epigenetic regulation of cartilage homeostasis has not been investigated, there is evidence in adult endothelial cells that shear stress regulates gene expression. Shear stress at 10 dyn/cm$^2$/sec modifies core histones H3 and H4 in human umbilical vein endothelial cells. It was hypothesized that this event may be required for shear stress-dependent expression of early response genes. Western blot analyses revealed that in endothelial cells exposed to shear stress, both histone H3 acetylation and phosphorylation were increased.[165] A subsequent study found similar shear stress-induced histone modifications in mouse embryonic stem cells.[166]

## Conclusions and perspectives

While overloading and underloading cause cartilage degradation, moderate loading is beneficial for healthy joints and evidence suggests may serve as a means of chondroprotection, especially when applied at a local level. In addition to the use of physiological joint loading in treating arthritis, LIPUS may emerge as a safe and noninvasive therapeutic alternative.

RA and OA may differ in origin and level of synovial inflammation, but in both cases, the synovium contributes inflammatory cytokines and MMP production into the joint fluid, and a common consequence for both diseases is cartilage destruction. Therefore, strategies to prevent cartilage degradation in arthritis should consider other joint tissues in addition to the articular cartilage.

Genetic regulation is important for cartilage health, but emerging data suggest that epigenetics play a critical role in cartilage homeostasis and the progression of OA and RA. The elucidation of these mechanisms may reveal novel therapeutic strategies in treating cartilage disease.

## Acknowledgments

The author would like to thank Daniel Leong for discussion and help in the preparation of the manuscript, and Dr. Luis Cardoso and Dr. Robert J. Majeska for critically reading the manuscript.

## Conflicts of interest

The author declares no conflicts of interest.

## References

1. Arendt, E. & R. Dick. 1995. Knee injury patterns among men and women in collegiate basketball and soccer. NCAA data and review of literature. *Am. J. Sports Med.* **23:** 694–701.

2. Jones, S.J., R.A. Lyons, J. Sibert, R. Evans, *et al.* 2001. Changes in sports injuries to children between 1983 and 1998: comparison of case series. *J. Public Health Med.* **23:** 268–271.

3. Levy, A.S., J. Lohnes, S. Sculley, M. LeCroy, *et al.* 1996. Chondral delamination of the knee in soccer players. *Am. J. Sports Med.* **24:** 634–639.

4. Mandelbaum, B.R. *et al.* 1998. Articular cartilage lesions of the knee. *Am. J. Sports Med.* **26:** 853–861.

5. Moti, A.W. & L.J. Micheli. 2003. Meniscal and articular cartilage injury in the skeletally immature knee. *Instr. Course Lect.* **52:** 683–690.

6. Smith, A.D. & S.S. Tao. 1995. Knee injuries in young athletes. *Clin. Sports Med.* **14:** 629–650.

7. Piasecki, D.P., K.P. Spindler, T.A. Warren, *et al.* 2003. Intraarticular injuries associated with anterior cruciate ligament tear: findings at ligament reconstruction in high school and recreational athletes. An analysis of sex-based differences. *Am. J. Sports Med.* **31:** 601–605.

8. Buckwalter, J.A. & J.A. Martin. 2006. Osteoarthritis. *Adv. Drug Deliv. Rev.* **58:** 150–167.

9. Saxon, L., C. Finch & S. Bass. 1999. Sports participation, sports injuries and osteoarthritis: implications for prevention. *Sports Med.* **28:** 123–135.

10. Torzilli, P.A., R. Grigiene, J. Borrelli Jr. & D.L. Helfet. 1999. Effect of impact load on articular cartilage: cell metabolism and viability, and matrix water content. *J. Biomech. Eng.* **121:** 433–441.

11. Loening, A.M. *et al.* 2000. Injurious mechanical compression of bovine articular cartilage induces chondrocyte apoptosis. *Arch. Biochem. Biophys.* **381:** 205–212.

12. Kurz, B. *et al.* 2001. Biosynthetic response and mechanical properties of articular cartilage after injurious compression. *J. Orthop. Res.* **19:** 1140–1146.

13. Morel, V. & T.M. Quinn. 2004. Cartilage injury by ramp compression near the gel diffusion rate. *J. Orthop. Res.* **22:** 145–151.

14. Chen, C.T., N. Burton-Wurster, G. Lust, R.A. Bank, *et al.* 1999. J.M. Compositional and metabolic changes in damaged cartilage are peak-stress, stress-rate, and loading-duration dependent. *J. Orthop. Res.* **17:** 870–879.

15. Chen, C.T. *et al.* 2001. Chondrocyte necrosis and apoptosis in impact damaged articular cartilage. *J. Orthop. Res.* **19:** 703–711.

16. Jones, M.H. & A.S. Amendola. 2007. Acute treatment of inversion ankle sprains: immobilization versus functional treatment. *Clin. Orthop. Relat. Res.* **455:** 169–172.

17. McCarthy, C. & E. Oakley. 2002. Management of suspected cervical spine injuries: the paediatric perspective. *Accid. Emerg. Nurs.* **10:** 163–169.

18. Fontaine, K.R., M. Heo & J. Bathon. 2004. Are US adults with arthritis meeting public health recommendations for physical activity? *Arthritis Rheum.* **50:** 624–628.

19. Haapala, J. *et al.* 1999. Remobilization does not fully restore immobilization induced articular cartilage atrophy. *Clin. Orthop. Relat. Res.* **362:** 218–229.

20. Jurvelin, J., I. Kiviranta, M. Tammi & J.H. Helminen. 1986. Softening of canine articular cartilage after immobilization of the knee joint. *Clin .Orthop. Relat. Res.* **207:** 246–252.

21. Haapala, J. *et al.* 2000. Incomplete restoration of immobilization induced softening of young beagle knee articular

cartilage after 50-week remobilization. *Int. J. Sports Med.* **21:** 76–81.

22. Haapala, J. *et al.* 1996. Coordinated regulation of hyaluronan and aggrecan content in the articular cartilage of immobilized and exercised dogs. *J. Rheumatol.* **23:** 1586–1593.

23. Evans, E.B., G.W.N. Eggers, J.K. Butler & J. Blumel. 1960. Experimental immobilization and remobilization of rat knee joints. *J. Bone Joint Surg. Am.* **42:** 737–758.

24. Hagiwara, Y. *et al.* 2009. Changes of articular cartilage after immobilization in a rat knee contracture model. *J. Orthop. Res.* **27:** 236–242.

25. Vanwanseele, B., F. Eckstein, H. Knecht, A. Spaepen. 2003. Longitudinal analysis of cartilage atrophy in the knees of patients with spinal cord injury. *Arthritis Rheum.* **48:** 3377–3381.

26. Hinterwimmer, S. *et al.* 2004. Cartilage atrophy in the knees of patients after seven weeks of partial load bearing. *Arthritis Rheum.* **50:** 2516–2520.

27. Buckwalter, J.A. & H.J. Mankin. 1998. Articular cartilage: degeneration and osteoarthritis, repair, regeneration, and transplantation. *Instr. Course Lect.* **47:** 487–504.

28. Jackson, D.W., P.A. Lalor, H.M. Aberman & T.M. Simon. 2001. Spontaneous repair of full-thickness defects of articular cartilage in a goat model. A preliminary study. *J. Bone Joint Surg. Am.* **83-A,** 53–64.

29. Vrahas, M.S., K. Mithoefer & D. Joseph. 2004. The long-term effects of articular impaction. *Clin. Orthop. Relat. Res.* **423:** 40–43.

30. Lohmander, L.S., H. Roos, L. Dahlberg, L.A. Hoerrner, *et al.* 1994. M.W. Temporal patterns of stromelysin-1, tissue inhibitor, and proteoglycan fragments in human knee joint fluid after injury to the cruciate ligament or meniscus. *J. Orthop. Res.* **12:** 21–28.

31. Peat, G., P. Croft & E. Hay. 2001. Clinical assessment of the osteoarthritis patient. *Best Pract. Res. Clin. Rheumatol.* **15:** 527–544.

32. Sinkov, V. & T. Cymet. 2003. Osteoarthritis: understanding the pathophysiology, genetics, and treatments. *J. Natl. Med. Assoc.* **95:** 475–482.

33. Hinton, R., R.L. Moody, A.W. Davis & S.F. Thomas. 2002. Osteoarthritis: diagnosis and therapeutic considerations. *Am. Fam. Physician.* **65:** 841–848.

34. Felson, D.T. *et al.* 2000. Osteoarthritis: new insights. Part 1: the disease and its risk factors. *Ann. Intern. Med.* **133:** 635–646.

35. Benito, M.J., D.J. Veale, O. FitzGerald & W.B. Van Den Berg. 2005. Synovial tissue inflammation in early and late osteoarthritis. *Ann. Rheum. Dis.* **64:** 1263–1267.

36. Abramson, S.B. & M. Attur. 2009. Developments in the scientific understanding of osteoarthritis. *Arthritis Res. Ther.* **11:** 227.

37. Derfus, B.A. *et al.* 2002. The high prevalence of pathologic calcium crystals in pre-operative knees. *J. Rheumatol.* **29:** 570–574.

38. Myers, S.L., D. Flusser, K.D. Brandt & D.A. Heck. 1992. Prevalence of cartilage shards in synovium and their association with synovitis in patients with early and endstage osteoarthritis. *J. Rheumatol.* **19:** 1247–1251.

39. Sutton, S. *et al.* 2009. The contribution of the synovium, synovial derived inflammatory cytokines and neuropeptides to the pathogenesis of osteoarthritis. *Vet. J.* **179:** 10–24.

40. Rollin, R. *et al.* 2008. Early lymphocyte activation in the synovial microenvironment in patients with osteoarthritis: comparison with rheumatoid arthritis patients and healthy controls. *Rheumatol. Int.* **28:** 757–764.

41. Da, R.R., Y. Qin, D. Baeten & Y. Zhang. 2007. B cell clonal expansion and somatic hypermutation of Ig variable heavy chain genes in the synovial membrane of patients with osteoarthritis. *J. Immunol.* **178:** 557–565.

42. Attur, M., J. Samuels, S. Krasnokutsky & S.B. Abramson. 2010. Targeting the synovial tissue for treating osteoarthritis (OA): where is the evidence? *Best Pract. Res. Clin. Rheumatol.* **24:** 71–79.

43. Abdel-Nasser, A.M., J.J. Rasker & H.A. Valkenburg. 1997. Epidemiological and clinical aspects relating to the variability of rheumatoid arthritis. *Semin. Arthritis Rheum.* **27:** 123–140.

44. Yelin, E., R. Meenan, M. Nevitt & W. Epstein. 1980. Work disability in rheumatoid arthritis: effects of disease, social, and work factors. *Ann. Intern. Med.* **93:** 551–556.

45. Kobelt, G., K. Eberhardt, L. Jonsson & B. Jonsson. 1999. Economic consequences of the progression of rheumatoid arthritis in Sweden. *Arthritis Rheum.* **42:** 347–356.

46. Arend, W.P. 2001. Cytokine imbalance in the pathogenesis of rheumatoid arthritis: the role of interleukin-1 receptor antagonist. *Semin. Arthritis Rheum.* **30:** 1–6.

47. Brennan, F.M. & I.B. McInnes. 2008. Evidence that cytokines play a role in rheumatoid arthritis. *J. Clin. Invest.* **118:** 3537–3545.

48. Tran, C.N., S.K. Lundy & D.A. Fox. 2005. Synovial biology and T cells in rheumatoid arthritis. *Pathophysiology.* **12:** 183–189.

49. Ferrara, N., H.P. Gerber & J. LeCouter. 2003. The biology of VEGF and its receptors. *Nat. Med.* **9:** 669–676.

50. Koch, A.E. 1998. Review: angiogenesis: implications for rheumatoid arthritis. *Arthritis Rheum.* **41:** 951–962.

51. Yoo, S.A., S.K. Kwok & W.U. Kim. 2008. Proinflammatory role of vascular endothelial growth factor in the pathogenesis of rheumatoid arthritis: prospects for therapeutic intervention. *Mediators Inflamm.* **2008:** 129873.

52. Lee, S.S. *et al.* 2001. Vascular endothelial growth factor levels in the serum and synovial fluid of patients with rheumatoid arthritis. *Clin. Exp. Rheumatol.* **19:** 321–324.

53. Mould, A.W. *et al.* 2003. Vegfb gene knockout mice display reduced pathology and synovial angiogenesis in both antigen-induced and collagen-induced models of arthritis. *Arthritis Rheum.* **48:** 2660–2669.

54. Goldring, M.B. & K.B. Marcu. 2009. Cartilage homeostasis in health and rheumatic diseases. *Arthritis Res. Ther.* **11:** 224–240.

55. Aigner, T. & L. McKenna. 2002. Molecular pathology and pathobiology of osteoarthritic cartilage. *Cell Mol. Life Sci.* **59:** 5–18.

56. Cawston, T.E. & A.J. Wilson. 2006. Understanding the role of tissue degrading enzymes and their inhibitors in

development and disease. *Best Pract. Res. Clin. Rheumatol.* **20:** 983–1002.

57. Rengel, Y., C. Ospelt & S. Gay. 2007. Proteinases in the joint: clinical relevance of proteinases in joint destruction. *Arthritis Res. Ther.* **9:** 221–230.

58. Burrage, P.S., K.S. Mix & C.E. Brinckerhoff. 2006. Matrix metalloproteinases: role in arthritis. *Front. Biosci.* **11:** 529–543.

59. Tortorella, M.D. *et al.* 1999. Purification and cloning of aggrecanase-1: a member of the ADAMTS family of proteins. *Science* **284:** 1664–1666.

60. Gendron, C. *et al.* 2007. Proteolytic activities of human ADAMTS-5: comparative studies with ADAMTS-4. *J. Biol. Chem.* **282:** 18294–18306.

61. Porter, S., I.M. Clark, L. Kevorkian & D.R. Edwards. 2005. The ADAMTS metalloproteinases. *Biochem. J.* **386:** 15–27.

62. Black, R.A. *et al.* 1997. A metalloproteinase disintegrin that releases tumour-necrosis factor-alpha from cells. *Nature* **385:** 729–733.

63. Moss, M.L. *et al.* 1997. Cloning of a disintegrin metalloproteinase that processes precursor tumour-necrosis factor-alpha. *Nature* **385:** 733–736.

64. Crowe, P.D. *et al.* 1995. A metalloprotease inhibitor blocks shedding of the 80-kD TNF receptor and TNF processing in T lymphocytes. *J. Exp. Med.* **181:** 1205–1210.

65. White, J.M. 2003. ADAMs: modulators of cell-cell and cell-matrix interactions. *Curr. Opin. Cell Biol.* **15:** 598–606.

66. Brew, K. & H. Nagase. 2010. The tissue inhibitors of metalloproteinases (TIMPs): an ancient family with structural and functional diversity. *Biochim. Biophys. Acta.* **1803:** 55–71.

67. Hamze, A.B. *et al.* 2007. Constraining specificity in the N-domain of tissue inhibitor of metalloproteinases-1: gelatinase-selective inhibitors. *Protein Sci.* **16:** 1905–1913.

68. Nagase, H. & G. Murphy. 2008. *Tailoring TIMPs for selective metalloproteinase inhibition.* Springer Science. New York.

69. Amour, A. *et al.* 2000. The in vitro activity of ADAM-10 is inhibited by TIMP-1 and TIMP-3. *FEBS Lett.* **473:** 275–279.

70. Amour, A. *et al.* 1998. TNF-alpha converting enzyme (TACE) is inhibited by TIMP-3. *FEBS Lett.* **435:** 39–44.

71. Kashiwagi, M., M. Tortorella, H. Nagase & K. Brew. 2001. TIMP-3 is a potent inhibitor of aggrecanase 1 (ADAM-TS4) and aggrecanase 2 (ADAM-TS5). *J. Biol. Chem.* **276:** 12501–12504.

72. Wang, W.M., G. Ge, N.H. Lim, *et al.* 2006. TIMP-3 inhibits the procollagen N-proteinase ADAMTS-2. *Biochem. J.* **398:** 515–519.

73. Sahebjam, S., R. Khokha & J.S. Mort. 2007. Increased collagen and aggrecan degradation with age in the joints of Timp3(-/-) mice. *Arthritis Rheum.* **56:** 905–909.

74. Fanjul-Fernandez, M., A.R. Folgueras, S. Cabrera & C. Lopez-Otin. 2010. Matrix metalloproteinases: evolution, gene regulation and functional analysis in mouse models. *Biochim. Biophys. Acta.* **1803:** 3–19.

75. McQuibban, G.A. *et al.* 2002. Matrix metalloproteinase processing of monocyte chemoattractant proteins generates CC chemokine receptor antagonists with anti-inflammatory properties in vivo. *Blood* **100:** 1160–1167.

76. Nagase, H., R. Visse & G. Murphy. 2006. Structure and function of matrix metalloproteinases and TIMPs. *Cardiovasc. Res.* **69:** 562–573.

77. Itoh, T. *et al.* 2002. The role of matrix metalloproteinase-2 and matrix metalloproteinase-9 in antibody-induced arthritis. *J. Immunol.* **169:** 2643–2647.

78. Clements, K.M. *et al.* 2003. Gene deletion of either interleukin-1beta, interleukin-1beta-converting enzyme, inducible nitric oxide synthase, or stromelysin 1 accelerates the development of knee osteoarthritis in mice after surgical transection of the medial collateral ligament and partial medial meniscectomy. *Arthritis Rheum.* **48:** 3452–3463.

79. Holmbeck, K. *et al.* 1999. MT1-MMP-deficient mice develop dwarfism, osteopenia, arthritis, and connective tissue disease due to inadequate collagen turnover. *Cell* **99:** 81–92.

80. Bohm, B.B. *et al.* 2005. Homeostatic effects of the metalloproteinase disintegrin ADAM15 in degenerative cartilage remodeling. *Arthritis Rheum.* **52:** 1100–1109.

81. Ferretti, M. *et al.* 2006. Biomechanical signals suppress proinflammatory responses in cartilage: early events in experimental antigen-induced arthritis. *J. Immunol.* **177:** 8757–8766.

82. Ferretti, M. *et al.* 2005. Anti-inflammatory effects of continuous passive motion on meniscal fibrocartilage. *J Orthop Res.* **23:** 1165–1171.

83. Salter, R.B. 2004. Continuous passive motion: from origination to research to clinical applications. *J. Rheumatol.* **31:** 2104–2105.

84. Ikenoue, T. *et al.* 2003. Mechanoregulation of human articular chondrocyte aggrecan and type II collagen expression by intermittent hydrostatic pressure in vitro. *J. Orthop. Res.* **21:** 110–116.

85. Lee, D.A. & D.L. Bader. 1997. Compressive strains at physiological frequencies influence the metabolism of chondrocytes seeded in agarose. *J. Orthop. Res.* **15:** 181–188.

86. Shelton, J.C., D.L. Bader & D.A. Lee. 2003. Mechanical conditioning influences the metabolic response of cell-seeded constructs. *Cells Tissues Organs* **175:** 140–150.

87. Mauck, R.L. *et al.* 2000. Functional tissue engineering of articular cartilage through dynamic loading of chondrocyte-seeded agarose gels. *J. Biomech. Eng.* **122:** 252–260.

88. Sharma, G., R.K. Saxena & P. Mishra. 2007. Differential effects of cyclic and static pressure on biochemical and morphological properties of chondrocytes from articular cartilage. *Clin Biomech (Bristol, Avon).* **22:** 248–255.

89. Millward-Sadler, S.J., M.O. Wright, L.W. Davies, *et al.* 2000. Mechanotransduction via integrins and interleukin-4 results in altered aggrecan and matrix metalloproteinase 3 gene expression in normal, but not osteoarthritic, human articular chondrocytes. *Arthritis Rheum.* **43:** 2091–2099.

90. Kamiya, T. *et al.* 2010. Effects of mechanical stimuli on the synthesis of superficial zone protein in chondrocytes. *J. Biomed. Mater. Res. A.* **92:** 801–805.

91. Parkkinen, J.J., M.J. Lammi, H.J. Helminen & M. Tammi. 1992. Local stimulation of proteoglycan synthesis in articular cartilage explants by dynamic compression in vitro. *J. Orthop. Res.* **10:** 610–620.

92. Valhmu, W.B. *et al*. 1998. Load-controlled compression of articular cartilage induces a transient stimulation of aggrecan gene expression. *Arch. Biochem. Biophys.* **353:** 29–36.

93. Otterness, I.G. *et al*. 1998. Exercise protects against articular cartilage degeneration in the hamster. *Arthritis Rheum.* **41:** 2068–2076.

94. Galois, L. *et al*. 2003. Moderate-impact exercise is associated with decreased severity of experimental osteoarthritis in rats. *Rheumatology (Oxford)* **42:** 692–693. author reply 693–694.

95. Manninen, P., H. Riihimaki, M. Heliovaara & O. Suomalainen. 2001. Physical exercise and risk of severe knee osteoarthritis requiring arthroplasty. *Rheumatology (Oxford)* **40:** 432–437.

96. Roos, E.M. & L. Dahlberg. 2005. Positive effects of moderate exercise on glycosaminoglycan content in knee cartilage: a four-month, randomized, controlled trial in patients at risk of osteoarthritis. *Arthritis Rheum.* **52:** 3507–3514.

97. Miller, G.D. *et al*. 2003. The Arthritis, Diet and Activity Promotion Trial (ADAPT): design, rationale, and baseline results. *Control Clin. Trials* **24:** 462–480.

98. American College of Rheumatology Subcommittee on Osteoarthritis Guidelines. 2000. Recommendations for the medical management of osteoarthritis of the hip and knee: 2000 update. *Arthritis Rheum.* **43:** 1905–1915.

99. Jordan, K.M. *et al*. 2003. EULAR Recommendations 2003: an evidence based approach to the management of knee osteoarthritis: report of a Task Force of the Standing Committee for International Clinical Studies Including Therapeutic Trials (ESCISIT). *Ann. Rheum. Dis.* **62:** 1145–1155.

100. Ettinger, W.H. Jr., *et al*. 1997. A randomized trial comparing aerobic exercise and resistance exercise with a health education program in older adults with knee osteoarthritis. The Fitness Arthritis and Seniors Trial (FAST). *JAMA.* **277:** 25–31.

101. O'Reilly, S.C., K.R. Muir & M. Doherty. 1999. Effectiveness of home exercise on pain and disability from osteoarthritis of the knee: a randomised controlled trial. *Ann. Rheum. Dis.* **58:** 15–19.

102. van Baar, M.E. *et al*. 1998. The effectiveness of exercise therapy in patients with osteoarthritis of the hip or knee: a randomized clinical trial. *J. Rheumatol.* **25:** 2432–2439.

103. Ekblom, B., O. Lovgren, M. Alderin, *et al*. 1975. Effect of short-term physical training on patients with rheumatoid arthritis. a six-month follow-up study. *Scand. J. Rheumatol.* **4:** 87–91.

104. Stenstrom, C.H. & M.A. Minor. 2003. Evidence for the benefit of aerobic and strengthening exercise in rheumatoid arthritis. *Arthritis Rheum.* **49:** 428–434.

105. Sambajon, V.V., J.E. Cillo Jr., R.J. Gassner & M.J. Buckley. 2003. The effects of mechanical strain on synovial fibroblasts. *J. Oral Maxillofac. Surg.* **61:** 707–712.

106. Sun, H.B., R. Nalim & H. Yokota. 2003. Expression and activities of matrix metalloproteinases under oscillatory shear in IL-1-stimulated synovial cells. *Connect Tissue Res.* **44:** 42–49.

107. Sun, H.B. & H. Yokota. 2002. Reduction of cytokine-induced expression and activity of MMP-1 and MMP-13

108. Torzilli, P.A., M. Bhargava, S. Park & C.T. Chen. 2010. Mechanical load inhibits IL-1 induced matrix degradation in articular cartilage. *Osteoarthritis Cartilage.* **1:** 97–105.

109. Heckman, J.D., J.P. Ryaby, J. McCabe, *et al*. 1994. Acceleration of tibial fracture-healing by non-invasive, low-intensity pulsed ultrasound. *J. Bone Joint Surg. Am.* **76:** 26–34.

110. Kristiansen, T.K., J.P. Ryaby, J. McCabe, *et al*. 1997. Accelerated healing of distal radial fractures with the use of specific, low-intensity ultrasound. A multicenter, prospective, randomized, double-blind, placebo-controlled study. *J. Bone Joint Surg. Am.* **79:** 961–973.

111. Watanabe, Y., T. Matsushita, M. Bhandari, *et al*. 2010. Ultrasound for fracture healing: current evidence. *J. Orthop. Trauma* **24**(Suppl. 1): S56–S61.

112. Mukai, S. *et al*. 2005. Transforming growth factor-beta1 mediates the effects of low-intensity pulsed ultrasound in chondrocytes. *Ultrasound Med. Biol.* **31:** 1713–1721.

113. Naito, K. *et al*. 2010. Low-intensity pulsed ultrasound (LIPUS) increases the articular cartilage type II collagen in a rat osteoarthritis model. *J. Orthop. Res.* **28:** 361–369.

114. Rutjes, A.W., E. Nuesch, R. Sterchi & P. Juni. 2010. Therapeutic ultrasound for osteoarthritis of the knee or hip. *Cochrane Database Syst. Rev.* CD003132.

115. Lee, J.H., J.B. Fitzgerald, M.A. Dimicco & A.J. Grodzinsky. 2005. Mechanical injury of cartilage explants causes specific time-dependent changes in chondrocyte gene expression. *Arthritis Rheum.* **52:** 2386–2395.

116. Blain, E.J. 2007. Mechanical regulation of matrix metalloproteinases. *Front. Biosci.* **12:** 507–527.

117. Wong, M. & D.R. Carter. 2003. Articular cartilage functional histomorphology and mechanobiology: a research perspective. *Bone* **33:** 1–13.

118. Fitzgerald, J.B. *et al*. 2008. Shear- and compression-induced chondrocyte transcription requires MAPK activation in cartilage explants. *J. Biol. Chem.* **283:** 6735–6743.

119. Agarwal, S. *et al*. 2004. Role of NF-kappaB transcription factors in antiinflammatory and proinflammatory actions of mechanical signals. *Arthritis Rheum.* **50:** 3541–3548.

120. Pulai, J.I. *et al*. 2005. NF-kappa B mediates the stimulation of cytokine and chemokine expression by human articular chondrocytes in response to fibronectin fragments. *J. Immunol.* **174:** 5781–5788.

121. Pufe, T. *et al*. 2004. Vascular endothelial growth factor (VEGF) induces matrix metalloproteinase expression in immortalized chondrocytes. *J. Pathol.* **202:** 367–374.

122. Kurz, B. *et al*. 2005. Pathomechanisms of cartilage destruction by mechanical injury. *Ann. Anat.* **187:** 473–485.

123. Amin, A.R. *et al*. 1997. Superinduction of cyclooxygenase-2 activity in human osteoarthritis-affected cartilage. Influence of nitric oxide. *J. Clin. Invest.* **99:** 1231–1237.

124. Kang, R.Y., J. Freire-Moar, E. Sigal & C.Q. Chu. 1996. Expression of cyclooxygenase-2 in human and an animal model of rheumatoid arthritis. *Br. J. Rheumatol.* **35:** 711–718.

125. Sano, H. *et al*. 1992. In vivo cyclooxygenase expression in synovial tissues of patients with rheumatoid arthritis and

by mechanical strain in MH7A rheumatoid synovial cells. *Matrix Biol.* **21:** 263–270.

osteoarthritis and rats with adjuvant and streptococcal cell wall arthritis. *J. Clin. Invest*. **89**: 97–108.

126. Anderson, G.D. *et al*. 1996. Selective inhibition of cyclooxygenase (COX)-2 reverses inflammation and expression of COX-2 and interleukin 6 in rat adjuvant arthritis. *J. Clin. Invest*. **97**: 2672–2679.

127. Portanova, J.P. *et al*. 1996. Selective neutralization of prostaglandin E2 blocks inflammation, hyperalgesia, and interleukin 6 production in vivo. *J. Exp. Med*. **184**: 883–891.

128. Healy, Z.R., F. Zhu, J.D. Stull & K. Konstantopoulos. 2008. Elucidation of the signaling network of COX-2 induction in sheared chondrocytes: COX-2 is induced via a Rac/MEKK1/MKK7/JNK2/c-Jun-C/EBPbeta-dependent pathway. *Am. J. Physiol. Cell Physiol*. **294**: C1146–C1157.

129. Hardy, M.M. *et al*. 2002. Cyclooxygenase 2-dependent prostaglandin E2 modulates cartilage proteoglycan degradation in human osteoarthritis explants. *Arthritis Rheum*. **46**: 1789–1803.

130. Laufer, S. 2003. Role of eicosanoids in structural degradation in osteoarthritis. *Curr. Opin. Rheumatol*. **15**: 623–627.

131. Miwa, M. *et al*. 2000. Induction of apoptosis in bovine articular chondrocyte by prostaglandin E(2) through cAMP-dependent pathway. *Osteoarthritis Cartilage* **8**: 17–24.

132. Jacques, C. *et al*. 1999. Cyclooxygenase activity in chondrocytes from osteoarthritic and healthy cartilage. *Rev. Rhum. Engl. Ed*. **66**: 701–704.

133. McCoy, J.M., J.R. Wicks & L.P. Audoly. 2002. The role of prostaglandin E2 receptors in the pathogenesis of rheumatoid arthritis. *J. Clin. Invest*. **110**: 651–658.

134. Jeffrey, J.E. & R.M. Aspden. 2007. Cyclooxygenase inhibition lowers prostaglandin E2 release from articular cartilage and reduces apoptosis but not proteoglycan degradation following an impact load in vitro. *Arthritis Res. Ther*. **9**: R129–R139.

135. Chowdhury, T.T., D.M. Salter, D.L. Bader & D.A. Lee. 2008. Signal transduction pathways involving p38 MAPK, JNK, NFkappaB and AP-1 influences the response of chondrocytes cultured in agarose constructs to IL-1beta and dynamic compression. *Inflamm. Res*. **57**: 306–313.

136. Ando, A. *et al*. 2009. Increased expression of metalloproteinase-8 and -13 on articular cartilage in a rat immobilized knee model. *Tohoku J. Exp. Med*. **217**: 271–278.

137. Leong, D.J. *et al*. 2010. Matrix metalloproteinase-3 in articular cartilage is upregulated by joint immobilization and suppressed by passive joint motion. *Matrix Biol*. **29**: 420–426.

138. Vanwanseele, B., E. Lucchinetti & E. Stussi. 2002. The effects of immobilization on the characteristics of articular cartilage: current concepts and future directions. *Osteoarthritis Cartilage* **10**: 408–419.

139. Sakamoto, J. *et al*. 2009. Immobilization-induced cartilage degeneration mediated through expression of hypoxia-inducible factor-1alpha, vascular endothelial growth factor, and chondromodulin-I. *Connect Tissue Res*. **50**: 37–45.

140. Trudel, G., H.K. Uhthoff & O. Laneuville. 2005. Prothrombin gene expression in articular cartilage with a putative role in cartilage degeneration secondary to joint immobility. *J. Rheumatol*. **32**: 1547–1555.

141. Basso, N. & J.N. Heersche. 2006. Effects of hind limb unloading and reloading on nitric oxide synthase expression and apoptosis of osteocytes and chondrocytes. *Bone* **39**: 807–814.

142. Chowdhury, T.T., D.L. Bader & D.A. Lee. 2006. Anti-inflammatory effects of IL-4 and dynamic compression in IL-1beta stimulated chondrocytes. *Biochem. Biophys. Res. Commun*. **339**: 241–247.

143. Chowdhury, T.T., D.L. Bader & D.A. Lee. 2003. Dynamic compression counteracts IL-1 beta-induced release of nitric oxide and PGE2 by superficial zone chondrocytes cultured in agarose constructs. *Osteoarthritis Cartilage* **11**: 688–696.

144. Deschner, J., B. Rath-Deschner & S. Agarwal. 2006. Regulation of matrix metalloproteinase expression by dynamic tensile strain in rat fibrochondrocytes. *Osteoarthritis Cartilage* **14**: 264–272.

145. Yokota, H., M.B. Goldring & H.B. Sun. 2003. CITED2-mediated regulation of MMP-1 and MMP-13 in human chondrocytes under flow shear. *J. Biol. Chem*. **278**: 47275–47280.

146. Dittmer, J. 2003. The biology of the Ets1 proto-oncogene. *Mol. Cancer* **2**: 29–50

147. Millward-Sadler, S.J. *et al*. 1999. Integrin-regulated secretion of interleukin 4: a novel pathway of mechanotransduction in human articular chondrocytes. *J. Cell Biol*. **145**: 183–189.

148. Wright, M.O. *et al*. 1997. Hyperpolarisation of cultured human chondrocytes following cyclical pressure-induced strain: evidence of a role for alpha 5 beta 1 integrin as a chondrocyte mechanoreceptor. *J. Orthop. Res*. **15**: 742–747.

149. Seguin, C.A. & S.M. Bernier. 2003. TNFalpha suppresses link protein and type II collagen expression in chondrocytes: Role of MEK1/2 and NF-kappaB signaling pathways. *J. Cell Physiol*. **197**: 356–369.

150. Ghosh, S. & M. Karin. 2002. Missing pieces in the NF-kappaB puzzle. *Cell* **109** (Suppl.): S81–S96.

151. Hoffmann, A., A. Levchenko, M.L. Scott & D. Baltimore. 2002. The IkappaB-NF-kappaB signaling module: temporal control and selective gene activation. *Science* **298**: 1241–1245.

152. Liacini, A. *et al*. 2003. Induction of matrix metalloproteinase-13 gene expression by TNF-alpha is mediated by MAP kinases, AP-1, and NF-kappaB transcription factors in articular chondrocytes. *Exp. Cell. Res*. **288**: 208–217.

153. Madhavan, S. *et al*. 2007. Biomechanical signals suppress TAK1 activation to inhibit NF-kappaB transcriptional activation in fibrochondrocytes. *J. Immunol*. **179**: 6246–6254.

154. Dossumbekova, A. *et al*. 2007. Biomechanical signals inhibit IKK activity to attenuate NF-kappaB transcription activity in inflamed chondrocytes. *Arthritis Rheum*. **56**: 3284–3296.

155. Bird, A. 2002. DNA methylation patterns and epigenetic memory. *Genes Dev*. **16**: 6–21.

156. Reik, W. 2007. Stability and flexibility of epigenetic gene regulation in mammalian development. *Nature* **447**: 425–432.

157. Feinberg, A.P. 2007. Phenotypic plasticity and the epigenetics of human disease. *Nature* **447:** 433–440.

158. Cheung, K.S., K. Hashimoto, N. Yamada & H.I. Roach. 2009. Expression of ADAMTS-4 by chondrocytes in the surface zone of human osteoarthritic cartilage is regulated by epigenetic DNA de-methylation. *Rheumatol. Int.* **29:** 525–534.

159. H.I. Roach, *et al.* 2005. Association between the abnormal expression of matrix-degrading enzymes by human osteoarthritic chondrocytes and demethylation of specific CpG sites in the promoter regions. *Arthritis Rheum.* **52:** 3110–3124.

160. Iliopoulos, D., K.N. Malizos & A. Tsezou. 2007. Epigenetic regulation of leptin affects MMP-13 expression in osteoarthritic chondrocytes: possible molecular target for osteoarthritis therapeutic intervention. *Ann. Rheum. Dis.* **66:** 1616–1621.

161. Loeser, R.F., H.J. Im, B. Richardson, *et al.* 2009. Methylation of the OP-1 promoter: potential role in the age-related decline in OP-1 expression in cartilage. *Osteoarthritis Cartilage* **17:** 513–517.

162. Kouzarides, T. 2000. Acetylation: a regulatory modification to rival phosphorylation? *EMBO J.* **19:** 1176–1179.

163. Young, D.A. *et al.* 2005. Histone deacetylase inhibitors modulate metalloproteinase gene expression in chondrocytes and block cartilage resorption. *Arthritis Res. Ther.* **7:** R503–R512.

164. Chung, Y.L., M.Y. Lee, A.J. Wang & L.F. Yao. 2003. A therapeutic strategy uses histone deacetylase inhibitors to modulate the expression of genes involved in the pathogenesis of rheumatoid arthritis. *Mol. Ther.* **8:** 707–717.

165. Illi, B. *et al.* 2003. Shear stress-mediated chromatin remodeling provides molecular basis for flow-dependent regulation of gene expression. *Circ. Res.* **93:** 155–161.

166. Illi, B. *et al.* 2005. Epigenetic histone modification and cardiovascular lineage programming in mouse embryonic stem cells exposed to laminar shear stress. *Circ. Res.* **96:** 501–508.

Ann. N.Y. Acad. Sci. ISSN 0077-8923

# ANNALS OF THE NEW YORK ACADEMY OF SCIENCES
Issue: *Molecular and Integrative Physiology of the Musculoskeletal System*

# Bone loss in anorexia nervosa: leptin, serotonin, and the sympathetic nervous system

Kevin K. Kumar,[1] Stephanie Tung,[2] and Jameel Iqbal[3]

[1]Vanderbilt University School of Medicine, Nashville, Tennessee. [2]Columbia University Medical Center, New York, New York. [3]Department of Pathology, Hospital of the University of Pennsylvania, Philadelphia, Pennsylvania

Address for correspondence: Jameel Iqbal, M.D., Ph.D., Department of Pathology, Hospital of the University of Pennsylvania, 3400 Spruce Street, Founders Pavilion 6th Floor, Philadelphia, PA 19104. drjamesiqbal@gmail.com

Anorexia nervosa (AN), a disorder characterized by the refusal to sustain a healthy weight, has the highest mortality of any psychiatric disorder. This review presents a model of AN that ties together advances in our understanding of how leptin, serotonin, and hypogonadism are brought about in AN and how they influence bone mass. Serotonin (5-hydroxytryptamine) is a key regulator of satiety and mood. The primary disturbance in AN results from alterations in serotonin signaling. AN patients suffer from serotonergic hyperactivity of Htr1a-dependent pathways that causes dysphoric mood and promotes restrictive behavior. By limiting carbohydrate ingestion, anorexics decrease their serotonin levels. Reduced serotonergic signaling in turn suppresses appetite through Htr1a/2b, decreases dysphoric mood through Htr1a/2a, and activates the sympathetic nervous system (SNS) through Htr2c receptors in the ventromedial hypothalamus. Activation of the SNS decreases bone mass through $\beta$2-adrenergic signaling in osteoblasts. Additional topics reviewed here include osteoblastic feedback of metabolism in anorexia, mechanisms whereby dietary changes exacerbate bone loss, the role of caloric restriction and *Sirt1* in bone metabolism, hypothalamic hypogonadism's effects on bone mass, and potential treatments.

Keywords: anorexia nervosa; bone loss; serotonin; osteoblast; leptin

## Introduction

Anorexia nervosa (AN) is an eating disorder characterized by the refusal to sustain a healthy weight. Coupled with this refusal is a distorted self-image that leads to the intense fear of gaining weight. Both genetic and psychosocial factors play a role in the development of AN. Although the stereotypical AN patient is a young Caucasian woman, AN affects all sexes, ages, and cultural backgrounds.

The Diagnostic and Statistical Manual of Mental Disorders-IV (DSM-IV) criteria for a diagnosis of AN includes (1) an intense fear of gaining weight; (2) a refusal to keep body weight >85% expected for age/height; (3) amenorrhea for three consecutive months; and (4) psychosocial distortions in one's self-image that may include a refusal to admit the significance of being underweight.

AN has the highest mortality rate of any psychiatric disorder.[1] Because AN is associated with malnutrition, every major organ system is affected. Lanugo, joint swelling, and abdominal distension are notable features of the disease.[1,2] Other notable characteristics of AN include dental cavities and tooth loss.[3] Many of the endocrine and metabolic changes seen in AN are compensatory, and are hypothesized to stimulate food intake and/or conserve the body's resources. However, some hormonal changes compound the disease; these changes force AN patients into a vicious cycle where weight loss and malnutrition fuel the desire to further restrict their diet.[4]

The treatment of AN is clinically challenging because of a lack of clear understanding of the underlying neuropsychiatric changes. Unlike other mental illnesses, effective pharmacological interventions for AN have not been identified. Like other serious mental illnesses though, patients with AN resist treatment—they are frequently reluctant to restore their weight.[1] Therefore, the development of an

doi: 10.1111/j.1749-6632.2010.05810.x

Ann. N.Y. Acad. Sci. 1211 (2010) 51–65 © 2010 New York Academy of Sciences.

effective treatment for AN remains a high-priority area of research.

## Neurological alterations

Downstream targets of leptin have been shown to play a critical role in the body's compensatory response to caloric restriction (CR) caused by AN. AN patients display altered concentrations of leptin, neuropeptide Y (NPY), corticotropin-releasing hormone (CRH), cholecystokinin, β-endorphin, and pancreatic polypeptide, among many others.[4] In addition to their direct effects on hunger and satiety, these molecules alter mood, the autonomic nervous system, hormone secretion, and cognitive function.[5] For example, intracerebroventricular (ICV) CRH infusion can cause the hypothalamic hypogonadism, hyperactivity, and altered sexual and feeding behavior frequently observed in AN patients.[4]

There are many comorbidities that are exhibited by patients suffering from AN, such as inhibition, anxiety, depression, obsessive behavior, and anhedonia. These comorbidities are mediated by alterations in two neurocircuits: the ventral (limbic) neurocircuit and the dorsal (cognitive) neurocircuit.[6,7] The ventral neurocircuit is composed of the amygdala, ventral straium, insula, anterior cingulate cortex, and the orbitofrontal cortex. The dorsal neurocircuit includes the hippocampus, dorsolateral prefrontal cortex, parietal cortex, and the dorsal regions of the anterior cingulate cortex.[6,7] Evidence for alterations in these circuits comes from imaging studies in AN patients that have demonstrated changes in the activity of frontal, parietal, and anterior cingulate regions.[8–10]

## Role of leptin in the neurobiology of food intake

### Introduction to leptin

Leptin, a 167-amino acid peptide, is a hormone critical to the regulation of body fat and appetite.[11] It is not surprising then that leptin is altered in AN patients: they have dramatically decreased leptin levels, on average >6-fold lower than normal peer controls.[12] Although leptin is secreted from a number of different sites, such as bone marrow, lymphoid tissue, placenta, and ovaries,[13] serum leptin levels are predominantly influenced by the amount of adipose tissue. Adipocyte leptin secretion relays information about peripheral energy storage and availability to the brain.[14] Increased levels of leptin serve to inhibit food intake, whereas decreased levels stimulate it.[15,16] Given the aforementioned characteristics of leptin, it is initially surprising that AN patients with their low levels of leptin exhibit reduced food intake.

### Leptin receptors and intracellular signaling

Leptin binds to leptin receptors (*ObRs*), which exists in multiple isoforms (*ObRa, ObRb, ObRc, ObRd, ObRe,* and *ObRf*).[17] Leptin isoforms have distinct functions: *ObRa* and *ObRc* function to transport leptin across the blood–brain barrier, whereas *ObRb* is responsible for leptinergic signaling.[17–20]

Knockout mice for leptin (denoted *ob/ob*) or for its receptor (*db/db*) have an early onset of severe obesity.[16,21] Consistent with the role of leptin in reducing food intake, administration of leptin to *ob/ob* mice reverses their obese phenotype.[22–24] Unfortunately for obese patients, they appear irresponsive to leptin; in other words, obese individuals exhibit resistance to leptin's actions.[25]

Intracellular signaling by leptin through the *ObRb* is mediated by tyrosine kinase autophosphorylation. JAK–STAT signaling is induced downstream of *ObRb*.[26] Specifically, the phosphorylated intracellular domain of *ObRb* serves as a binding site for STATs, which subsequently act as transcription factors.[27] For example, leptin induces STAT3 activation, which then increases transcription of the appetite-inducing molecule proopiomelanocrotin (POMC).[28]

### Leptinergic pathways in the central nervous system (CNS)

Leptin suppresses appetite (energy intake) and promotes energy expenditure by regulating neuronal activity in multiple regions of the hypothalamus, including the arcuate nucleus (ARC), ventromedial hypothalamus (VMH), and paraventricular hypothalamus.[29]

One area of prominent leptin action is in the hypothalamic ARC, which is a neuronal population in the mediobasal hypothalamus (Fig. 1).[30–32] Restoration of leptin signaling exclusively in the ARC of leptin receptor-deficient mice can correct the obesity and excessive food intake of these animals.[29,33] The ARC is composed of two major neuronal classes that express the leptin receptor: neurons-expressing POMC, and neurons-expressing NPY and agouti-related protein (AgRP).

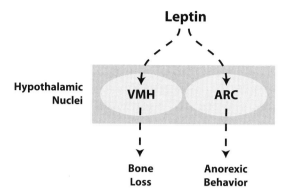

**Figure 1.** General overview of leptin action in the control of bone mass and feeding behavior. Leptin acts through indirect means on the ventromedial hypothalamus (VMH) to stimulate bone loss. Leptin acts through indirect means on the arcuate nucleus (ARC) to cause anorexic behavior. Figures 2 and 3 expound upon greater detail the mechanisms involved in anorexic behavior and bone loss, respectively.

POMC neurons inhibit food intake. These neurons coexpress two anorexigenic molecules: POMC and cocaine- and amphetamine-regulated transcript (CART).[29,33] Leptin infusion increases the expression of POMC and CART from these neurons.[34] Leptin and insulin both act to cause proteolytic cleavage of POMC to generate $\alpha$-melanocyte-stimulating hormone ($\alpha$-MSH).[35]

$\alpha$-MSH activates melanocortin-3 and melanocortin-4 receptors (MC3R and MC4R) to reduce food intake. For example, $\alpha$-MSH signaling through MC4R stimulates the central tegmental area and striatum; neurons in these areas alter mesolimbic dopamine signaling to inhibit food intake (Fig. 2).[36,37] The deletion of either MCR3 or MC4R results in an obese phenotype similar to leptin receptor-deficient mice.[35,38] Note that the deletion of both MC3R and MC4R results in a more severe obesity, suggesting that both receptors together normally mediate the anorexigenic effect.[29,39] Consistent with the function of MC4R in controlling food intake, the obese phenotype of MC3R and MC4R knockout mice can be reversed by restoration of MC4R in the paraventricular nucleus (PVN).[40]

In stark contrast to the POMC neurons, NPY/AgRP neuronal signaling acts to stimulate feeding.[41,42] NPY stimulates feeding, whereas AgRP antagonizes MC3R and MC4R, thereby blocking the anorexigenic actions of POMC neurons.[29] Selective ablation of AgRP neurons in adult mice causes extreme reduction in food intake, which results in death due to starvation.[43] Note that although leptin stimulates POMC and CART expression, leptin inhibits NPY and AgRP expression.[29] Thus, the overall actions of leptin in the ARC are to suppress signaling in neurons that stimulate food intake (NPY/AgRP) and stimulate signaling in neurons that reduce food intake (POMC).

There are several other mechanisms whereby leptin can control appetite. Leptin can modulate synaptic plasticity,[44] and through this mechanism it can promote the depolarization of POMC-expressing neurons while inhibiting those that express NPY and AgRP.[34] Leptin can also act as a neurotrophin during brain development; in this capacity, leptin is critical for the establishment of circuitry in the ARC as well as connections among the PVN, ventromedial nucleus (VMH), lateral hypothalamic area, and dorsomedial hypothalamic nucleus.[16] For example, it is known that *ObR* neurons in the VMH project to the ARC to modulate satiety by altering projection density;[45] indeed, genes that function to modulate synaptic plasticity—such as *Basign*, *Gap43*, *ApoE*, and *Gabrap*—are all positively regulated by leptin.[46]

Both AgRP and POMC neurons (first-order neurons) extensively innervate second-order neurons in the PVN. The PVN has also been shown to be critical to appetite regulation in that deletion of transcription factors vital for PVN development, such as *Sim1*, can result in obesity.[47–49] Alterations in PVN function are similar to many of the characteristics displayed in AN. For example, through the PVN leptin stimulates the sympathetic nervous system (SNS), thereby contributing to the promotion energy expenditure.[48] The PVN also contains neuroendocrine cells that secrete numerous anorexigenic factors, including thyrotropin-releasing hormone (TRH) and CRH, which are both stimulated by leptin via $\alpha$-MSH signaling. Note that AN patients have reduced TRH levels but increased CRH levels.[50,51]

Both AgRP and POMC neurons synapse on VMH neurons that express brain-derived neurotrophic factor (BDNF), another powerful anorexigenic molecule.[52] Leptin stimulates the expression of BDNF in the VMH through MC4R signaling.[52,53] Consistent with this, deletion of BDNF in the VMH/dorsomedial hypothalamus (DMH) of adult mice results in hyperphagia and obesity.[54]

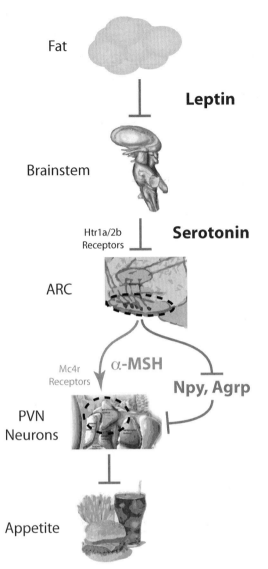

**Figure 2.** Pathways involved in leptin's control of appetite. Leptin acts through leptin receptors present on serotonergic neurons in the brainstem to inhibit serotonin synthesis and neuronal activity in an analogous fashion to its functions on bone metabolism. However, whereas the serotonergic neurons controlling bone metabolism project to the ventromedial hypothalamus (VMH), the serotonergic neurons controlling appetite project to the arcuate nucleus (ARC). In the ARC, serotonergic neurons act via Htr1a and 2b receptors, among others, to inhibit the function of POMC/CART neurons. MC4R signaling in the paraventricular nucleus (PVN) serves to inhibit appetite and food intake. In addition to effects on POMC neurons in the ARC, serotonin simultaneously inhibits the function of NPY and AGRP neurons in the ARC, which serve to increase appetite and inhibit MC4R neuronal signaling. Note that leptin likely directly acts on many different areas of the brain to mediate anorexigenic behavior; for example, leptin probably acts directly on the VMH to mediate BDNF production (not shown).

Similarly, deficiency of *TrkB*, the BDNF receptor, results in hyperphagia and obesity in both mice and humans.[55]

## Neurotransmitter alterations

### Role of serotonin in AN

The 5-hydroxytryptamine (5-HT) system is critical in regulating satiety, impulse control, and mood.[56–60] Restricting-type anorexics, those that lose weight by strictly regulating food intake, are affected by aberrations in serotonin homeostasis.[4] AN patients prior to the onset of their illness suffer a preanorexic phase characterized by overstimulation of their serotonin pathways resulting in excessive activation of 5-HT1A pathways.[4] It is thought that this increased 5-HT1A stimulation leads to a dysphoric mood characterized by an anxious, harm-avoidant temperament.[4]

Carbohydrate intake, through metabolic influences on tryptophan, increases brain-derived serotonin levels.[61,62] To suppress 5-HT-related dysphoria, AN patients starve themselves of carbohydrate, thereby lowering their serotonin levels.[63] In line with this, acutely ill AN patients have decreased cerebrospinal fluid (CSF) levels of 5-hydroxyindoleacetic acid (5-HIAA)—a marker of 5-HT concentrations.[64] Recuperation from their malnutrition dramatically increases levels of serotonin signaling in AN patients.[4] This recovery comes at the expense of extreme anxiety and dysphoria, however, and often necessitates hospitalization.

In addition to serotonin synthesis, 5-HT receptors are also abnormal in AN patients. Of the different serotonin-receptor subtypes, elevated 5-HT1A and decreased 5-HT2A receptors have been found in patients recovered from AN.[65–71] Note that these receptor subtypes are implicated in harm avoidance, a common AN trait consisting of anxiety, inhibition, and inflexibility.[72] Thus, serotonin pathways are thought to be integral to AN patient's aversion to eating.

Starvation or food restriction decreases extracellular 5-HT concentrations, thereby reducing stimulation of 5-HT1A and 5-HT2A postsynaptic receptors and associated dysphoria.[4] Conversely, when patients suffering from AN are forced to eat, 5-HT concentrations elevate and increase dysphoria. If starvation is prolonged, neuropeptides are modulated to exacerbate the symptoms of AN; for example, increases in CRH lead to further reduction in food intake and changes in behavior.[4]

Mice deficient in brain-derived serotonin, brought about through the genetic deletion of *Tph2* tryptophan hydroxylase ($Tph2^{-/-}$ mice), have reduced food intake despite normal levels of leptin, insulin, and corticosteroids.[73] This important finding suggests that leptin controls appetite, although also modulates serotonin levels (Fig. 2). Indeed, mice deficient in leptin have dramatically increased levels of serotonin.[73] Overall, the available evidence points to decreased serotonin as the key alteration causing appetite reduction in AN patients.

### Serotonin–dopamine interactions

Serotonin signaling in AN is closely linked to its interactions with dopamine, a neurotransmitter that modulates the reward pathway in the brain. Dysfunction in dopamine signaling, particularly in striatal circuits, contributes to alterations in the reward pathway consistent with the dietary restriction and impaired decision making present in AN.[74] These behavioral observations have physiologic correlates; individuals suffering or recovered from AN have reduced CSF dopamine metabolite levels and polymorphisms in its functional receptor DRD2.[75,76]

Serotonin has the capacity to modulate dopamine circuitry: 5-HT2C receptors tonically inhibit dopaminergic neurons.[77,78] In AN, unbalanced interactions between 5-HT and dopaminergic pathways causes altered interactions between dorsal and ventral neurocircuits.[4] These alterations are thought to reinforce some of the ritualistic food behaviors.

### Role of norepinephrine

Norepinephrine (NE) is known to influence food intake dependent on its target in the central nervous system (CNS).[79] Administration of NE to the PVN of the hypothalamus produces hyperphagia.[80] Conversely, administration of NE to the perifornical hypothalamus produces reduction in food intake.[80] This later effect has been used by the antiobesity drugs sibutramine and phentermine, which either block NE reuptake or induce NE release, respectively.[81]

Anorexics are often thought to have decreased activation of their SNS, which uses NE as its neurotransmitter. The evidence for this belief derives from studies demonstrating that AN patients who have sustained recovery of their weight display significantly lower levels of NE and metabolites in their

CSF and urine.[82,83] Although these observations suggest that there is a dysfunction in NE metabolism in AN patients, confusingly, normal levels of NE are found in the CSF of acutely ill AN patients.[84]

The theory that AN patients have lower SNS activity contrasts with several recent findings. First, if anorexics have a higher serotonin set point that then is lowered through food restriction, decreased serotonin levels should increase β-adrenergic stimulation (Fig. 3).[73] Second, a number of reports have demonstrated that AN patients have significantly increased levels of NE in their adipose tissue.[85,86]

In contrast to the aforementioned evidence of SNS overstimulation, many symptoms in acutely ill AN patients suggest they have SNS understimulation. We hypothesize that this may in part reflect a compensatory mechanism from SNS overstimulation. For example, SNS overstimulation is known to rapidly downregulate β-adrenergic receptors in many tissues;[87] interestingly, starvation on the order of 2–3 weeks leads to downregulation of β-adrenergic receptors as well.[88] Thus, it is possible that symptoms of SNS understimulation seen in AN patients may reflect compensation for initially increased SNS activity. Ultimately, this theory will need to be confirmed with further studies.

### Clinical evidence of bone loss

Osteopenia and osteoporosis are common long-term physical complications of AN.[89–91] Compared to normal women, those with AN have decreased bone mineral density (BMD).[91] Their skeletal mineral content is ∼25% less than that of aged-matched controls.[92] The decrease in bone mass is thought to be due to rapid bone loss, which occurs within 6 months of AN onset. Of the sites experiencing bone loss, the most notable are the lumbar spine, femoral neck, and radii.[92] Because of this significant bone loss, women with AN have three times the risk of fracture;[91] moreover, more than half of women with AN experience at least one fracture before 40 years of age.[91]

The bone loss of AN is typically characterized by high bone resorption markers and low bone formation markers.[93] This is significant in that the traditional high-turnover bone loss seen in hypogonadism (e.g., menopause) is characterized by increases in both bone resorption and formation markers.[94] Thus, in addition to changes causing

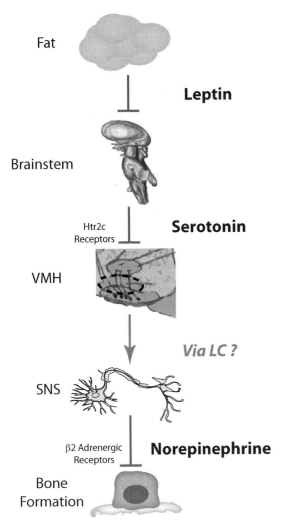

**Figure 3.** Pathways involved in leptin's control of bone metabolism. Leptin acts through leptin receptors present on serotonergic neurons in the brainstem to inhibit serotonin synthesis and neuronal activity. These serotonergic neurons project to the ventromedial hypothalamus (VMH) and inhibit the function of neurons there involved in the regulation of the sympathetic nervous system (SNS). It is likely that the VMH neurons mediate SNS function indirectly by regulating neuronal activity in the locus coeruleus, which are known to regulate SNS activity. The SNS has projections adjacent to osteoblasts in bone, and via activation of β2-adrenergic receptors can inhibit bone formation and stimulate bone resorption via upregulation of RANK-L.

increased resorption, AN patients also have alterations suppressing their bone formation.

Derivations in bone mass in AN can be attributed to more than just low body weight.[95] For example, in AN patients, decreases in BMD have been cor-

related with the duration of their amenorrhea.[96] Thus, other factors such as estrogen deficiency, glucocorticoid excesses, and malnutrition resulting in decreased vitamin and mineral intake are all thought to contribute towards osteopenia in AN patients.[97]

In evaluating the bone status in AN, during recovery bone resorption markers usually decrease, whereas those of bone formation invariably rise.[97,98] Unfortunately, the mineralization of newly formed bone in recovered AN patients is reduced; this occurs secondary to low intake of key vitamins and minerals such as $Ca^{2+}$ and vitamin $D_3$.[93] Low mineralization can often be reflected in BMD values during recovery. However, due to the young age of many AN patients, the use of $Z$-scores is recommended because these take into account age-expected increase in BMD values.[93]

## Mechanisms of bone loss

### Introduction to mechanisms of bone loss

Microcracks that can impair skeletal strength are constantly repaired with the removal of bone by osteoclasts and the subsequent deposition of collagen and mineral by osteoblasts. When resorption occurs in excess of formation, bone mass is reduced; if enough bone is degraded, then changes in skeletal architecture cause fragility and lead to fractures.

Osteoclast formation and function is governed by the cytokine receptor rctivator of NF-κB ligand (RANK-L). RANK-L is produced by several key cells: osteoblasts, bone lining cells, and T lymphocytes. Although the production of RANK-L from osteoblasts and bone lining cells helps to couple bone formation to bone degradation, osteoclasts dig up the bone matrix to release the cytokine TGFβ, which couples resorption to bone formation.

The intrabone resorption and formation cycles are subject to outside control from the CNS and endocrine systems. In this way, changes affecting an entire organism are translated into beneficial modifications of the bone resorption and formation cycle. Both the CNS and endocrine system play critical roles in mediating modifications of bone metabolism in AN.

### Leptin–serotonin alterations in bone loss

In 2000, Karsenty and colleagues discovered a role for leptin in the control of bone metabolism. Examining leptin-deficient *ob/ob* mice, they noted that these mice had increased bone mass compared to control mice.[99] Moreover, this phenotype was occurring despite *ob/ob* mice being hypogonadal and hypercortisolemic (the two most common causes of bone loss).[99] Using intercerebral infusion of leptin, they were able to demonstrate a reversal of the high bone mass phenotype—that is, leptin was acting in the CNS to cause bone loss.[99]

How is leptin controlling bone metabolism? As we have come to appreciate the role of serotonin and leptin in AN patients, this has been mirrored by an increased understanding of leptin action on bone metabolism and the role that serotonin plays in mediating the downstream effects of leptin. Examining the phenotype of mice deficient in brain-derived production of serotonin (BDS) has revealed the exact opposite phenotype from leptin-deficient mice.[73] These *Tph2*$^{-/-}$ mice have severely reduced bone mass and decreased bone formation and increased resorption parameters.[73] Note that the low bone mass phenotype of *Tph2*$^{-/-}$ mice occurred despite normal levels of leptin, insulin, and corticosteroids.[73] This suggests that, exactly like its role in controlling appetite, leptin regulates bone mass by modulating serotonin levels and not through changes in estrogen or corticosteroid levels (see Fig. 3).

How do alterations in BDS levels act to control bone mass? It is known that leptin deficiency results in a low sympathetic tone.[99] In contrast, BDS-deficient mice display characteristics of increased sympathetic output: they have increased levels of NE in the brain, epinephrine in the urine, and increased brown fat activity.[73] Genetic studies have confirmed that BDS acts on neurons in the VMH to decrease sympathetic activity.[73] Thus, in the absence of leptin, serotonin levels are high and the resulting sympathetic output is low. With leptin, however, serotonin levels decrease resulting in increased sympathetic output (see Fig. 3).

How does sympathetic output regulate bone metabolism? Genetic or pharmacological ablation of adrenergic signaling results in a high bone mass phenotype that is resistant to correction by intercerebral leptin infusion.[100] This suggests that adrenergic action on bone is downstream of leptin and serotonin. The activation of β-adrenergic receptors on osteoblasts decreases their proliferation; consistent with the notion that adrenergic action is downstream of leptin and serotonin, modulation of the β-adrenergic pathway in leptin-deficient mice can control bone mass.[100] For example, in leptin-deficient animals the addition of an adrenergic agonist decreases bone mass, whereas a β-adrenergic antagonist increases bone mass.[100]

Amazingly, none of the aforementioned adrenergic manipulations affected body weight.[100] This suggests that the leptin/serotonin pathways used in controlling appetite are distinct from those regulating bone mass. Indeed, the leptin-regulated

**Table 1.** Anorexogenic and bone metabolic phenotypes in knockout mice

| Feature | ObRb | Ob/Ob (leptin KO) | ObRb$^{-/-}$ Sf-1 VMH | ObRb$^{-/-}$ POMC ARC | ObRb$^{-/-}$ Sertonin neurons | Tphr$^{-/-}$ (Serotonin KO) | Htr2c$^{-/-}$ | AdrB2$^{-/-}$ | Htr2c$^{-/-}$ Htr2c$^{+/+}$ in VMH | CART$^{-/-}$ | Mcr4$^{-/-}$ |
|---|---|---|---|---|---|---|---|---|---|---|---|
| Sympathetic activity | ↓ | ↓ | Normal | Normal | ↓ | ↑ | ↑ | ↓ | Normal | ? | ? |
| Bone remodeling | BF ↑; BR ↑ | BF ↑; BR ↑ | Normal | Normal | ? | BF ↓; BR ↑ | BF ↓; BR ↑ | BF ↑; BR ↓ | Normal | BF normal; BR ↑ | BF normal; BR ↓ |
| Bone mass | ↑ | ↑ | Normal | Normal | ↑ | ↓ | ↓ | ↑ | Normal | ↓ | Late ↑ |
| Appetite | ↑ | ↑ | Normal | Normal | ↑ | ↓ | Late ↑ | Normal | ? | ? | ↑ |
| Body weight | ↑ | ↑ | Mild ↑ | Normal | ↑ | ↓ | Late ↑ | Normal | ? | On high fat ↑ | ↑ |
| Energy expenditure | ↓ | ↓ | Normal | Normal | ↓ | ? | ? | ? | ? | ? | ↓ |
| Brainstem serotonin | ↑ | ↑ | Normal | Normal | ↑ | N/A | ? | ? | ? | ? | ? |
| CART levels | ↓ | ↑ | ? | ? | ? | ? | ? | ? | ? | N/A | ↑ |

*Note*: In comparing the various phenotypes, whenever there is decreased sympathetic activity, there is increased bone mass; conversely, whenever sympathetic activity is increased, there is reduced bone mass. Therefore, SNS output to regulate bone metabolism is the common final effecter of hormonal and neuronal changes.

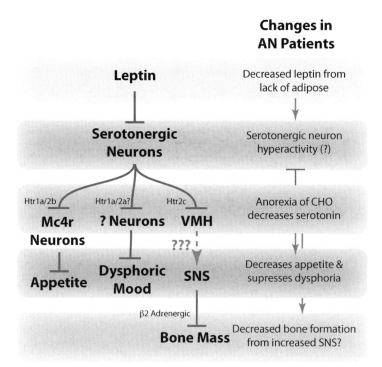

**Figure 4.** Overview of anorexia-induced alterations in neuronal signaling and their connections to appetite, dysphoric mood, and bone mass. Leptin is known to inhibit serotonergic neuronal activation, thereby decreasing levels of serotonin. Anorexia nervosa (AN) patients, however, have low levels of both leptin and serotonin. This reflects the thought that the primary disturbance in AN results from alterations in serotonin signaling, such that AN patients suffer from serotonergic hyperactivity of Htr1a-dependent pathways. Through restricting their carbohydrate ingestion, AN patients lower their levels of serotonin, thereby reducing the (hyper-)activity of serotonergic neurons. Serotonin signaling through Htr1a/2b pathways is known to regulate MC4R neurons; these later neurons serve to suppress appetite. Thus, the loss of serotonin in anorexia causes unchecked MC4R neuronal activity and decreased appetite. Serotonin, possibly through Htr1a/2a receptors, also functions to inhibit pathways that in turn inhibit dysphoric mood. Lastly, serotonin acts through Htr2c receptors to inhibit neurons in the ventromedial hypothalamus (VMH) that activate the SNS. The SNS, through β2-adrenergic receptors on osteoblasts, decreases bone mass. It should be noted that this last arm of serotonin function awaits confirmation in AN patients.

serotonergic neurons mediating bone mass preferentially project to the VMH, whereas the serotonergic neurons involved in appetite project to the ARC (Fig. 1). Consistent with these differential projections, distinctive sets of serotonin receptors mediate these effects.[73]

What can be gleaned about the mechanism of AN-induced bone loss from the leptin/serotonin/adrenergic studies? Table 1 summarizes the anorexogenic and bone phenotypes of numerous knockout mice. As stated in preceding sections, the phenotype of $Tph2^{-/-}$ mice is remarkably similar to part of the AN patient's phenotype. Knocking-out BDS resulted in decreased food intake and increased energy expenditure,[73] reminiscent of the behavioral aspects of AN.[4] Moreover, mice deficient in BDS have reduced levels of MC4R, thereby

suggesting that serotonin is altering appetite and energy expenditure in part through melanocortin signaling.[73]

The neural mechanisms of anorexia-induced changes are summarized in Figure 4. One of the treatments commonly used for AN are SSRIs. However, selective serotonin reuptake inhibitors (SSRIs) have been linked to causing bone loss themselves, and mice deficient in 5-hydroxytryptamine transporter (5-HTT) display low bone mass.[101] In this regard, it is unclear if the clinical effects of SSRIs are brought about through decreasing serotonergic signaling via receptor downregulation, or augmenting it through increased synaptic levels of serotonin. Regardless, the potential remains to modulate anorexia-associated or bone metabolism-associated serotonergic pathways (Fig. 4).

The role of adrenergic modulation as a potential treatment for AN-induced bone loss remains unknown. Theoretically, adrenergic blockade should help reverse the bone loss seen in AN patients, as well as some of the anxiety experienced by these individuals upon food consumption. However, as mentioned in the above sections on AN, many AN patients have signs of SNS understimulation making the administration of SNS blocking agents problematic.

### Follicle-stimulating hormone and estrogen in leptin-deficiency and AN-induced bone loss

Leptin-deficient mice are hypogonadal, yet have preserved bone mass. New research in hypogonadal bone loss has suggested that the hormone FSH plays a critical role in inducing osteoclast-driven bone breakdown;[94,102,103] FSH both directly and indirectly stimulates osteoclast activity to cause bone loss.[102,103] Leptin-deficient mice have no defects in their hematopoetic osteoclast progenitor cell potential; *ex vivo* osteoclast formation from *ob/ob* mice is normal.[99] However, *ob/ob* mice do have reduced levels of FSH.[104] Together, the above data suggest that osteoclastic derangements in hypogonadism are likely blunted in leptin-deficient animals by simultaneous decreases in their circulating FSH levels (Fig. 5).

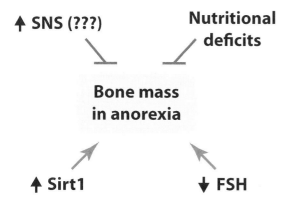

**Figure 5.** Multiple conflicting determinants of bone mass in anorexia. There are multiple determinants favoring both increases and decreases in bone mass in anorexia. Both increased sympathetic nervous system (SNS) activation and nutritional deficits (such as calcium and vitamin K) favor decreased bone mass. However, caloric restriction-induced *Sirt1* activity and decreased levels of follicle-stimulating hormone (FSH) both favor increases in bone mass. The overall outcome in AN is determined by the interplay of these various factors, but overall it favors bone loss.

The clinical implications of these findings are that osteoclastic derangements in AN are more likely to mirror changes in circulating FSH than estrogen. For example, in amenorrheic women with similar estrogen levels, having a mean serum FSH of 35 mIU/mL was correlated with considerably greater bone loss than those with a mean serum FSH level of 8 mIU/mL.[105]

### Persistent caloric deprivation and bone mass

Anorexic patients undergo persistent CR. Although one would anticipate that CR might decrease bone mass, the exact opposite is the case: CR increases bone mass. The mechanism whereby CR increases bone mass involves increasing osteoblast numbers while decreasing osteoclast numbers.[106] These effects have been postulated to occur through upregulation of Sirtuin 1 (*Sirt1*). In line with this reasoning, *sirt1* knockout mice display reduced bone mass.[106] The effects of CR on bone mass occur through *Sirt1* upregulation: mice deficient in *Sirt1* do not demonstrate increased bone mass upon CR.[106]

Although both the osteoblast- and osteoclast-specific *Sirt1* knockout mice display osteoporotic phenotypes,[106] the role of hypothalamic *Sirt1* in bone metabolism remains to be investigated. Interestingly, CR appears to increase *Sirt1* in POMC neurons, with the resulting effect of increasing appetite.[107] Loss of *Sirt1* expression in POMC neurons results in the inability of leptin to decrease appetite;[108] moreover, these animals have a hyperactive SNS,[108] hinting at the possibility that they may have reduced bone mass.

The clinical implications of the scientific studies on CR and bone mass suggest that the lack of food and body weight is not the cause of low bone mass. If anything, CR in anorexics should help prop-up bone mass (Fig. 5).

### Osteoblastic feedback on metabolism

Osteocalcin is a protein secreted by osteoblasts, and is generally considered a marker of their activity. AN patients have decreased osteoblast activity and thus reduced osteocalcin levels. However, there is new evidence that this relationship is not so simple, and that osteocalcin expression is controlled through neural mechanisms. Leptin knockout (*ob/ob*) mice have increased bone mass, yet their basal osteocalcin levels are significantly lower compared to controls.[109] Restoration of leptin expression via gene therapy can raise osteocalcin levels to normal.[109]

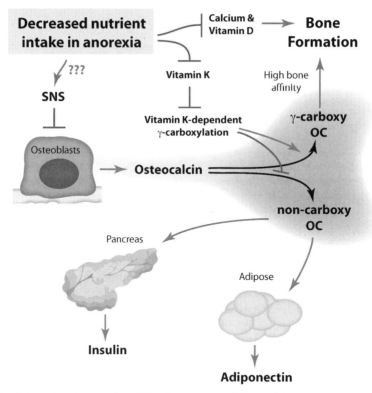

**Figure 6.** Nutritional alternations in anorexia and their impact on metabolism and bone formation. In anorexia nervosa (AN), decreased nutrient intake has several prominent effects: it lowers levels of serotonin in the brain, it decreases calcium and active vitamin D levels, and reduces vitamin K concentrations. Decreased serotonin is known to lead to enhanced β2-adrenergic stimulation, which in turn inhibits osteoblast function. Decreased osteoblast function can be observed as lowered levels of osteocalcin, a traditional marker of osteoblast-mediated bone formation. Interestingly, decreased activity of vitamin K-dependent enzymes can reduce the γ-carboxylation of osteocalcin. γ-Carboxylation is critical for increasing the affinity of osteocalcin for the bone. Noncarboxylated osteocalcin can act on the pancreas and adipose tissues to induce production of insulin and adiponectin, respectively. Note that AN patients display reduced levels of osteocalcin, and increased levels of insulin and adiponectin; the changes in γ-carboxylation of osteocalcin in AN remains to be determined. A lack of γ-carboxylated osteocalcin, combined with decreased calcium, results in significantly impaired bone formation and mineralization. AN patients typically suffer from both impaired bone formation and decreased mineralization.

These findings demonstrate that leptin has a surprisingly beneficial effect on osteoblast function, inducing osteocalcin efflux.

One reason for a connection between leptin and osteocalcin is that circulating osteocalcin may modulate insulin and weight homeostasis (Fig. 6).[109] An important notion from animal models is that an overall decrease in osteocalcin levels may not adequately reflect the osteocalcin biology without specific measurements of uncarboxylated osteocalcin. For example, despite a reduction in total serum levels of osteocalcin in *Igfbp2*[−/−] mice, there is actually increased levels of uncarboxylated osteocalcin.[110] Thus, it is possible that the decreased osteocalcin levels seen in AN patients may still have increased activity, albeit for the uncarboxylated form that acts on pancreas and fat, and not bone (Fig. 6). In support of this notion, both insulin and adiponectin—two hormones whose production is stimulated by uncarboxylated osteocalcin—are increased in AN patients.[12]

## Therapies for bone loss

### Nutritional supplementation

During a state of starvation, serum calcium homeostasis takes precedence over maintenance of bone tissue.[92] Thus, women with AN tend to be in negative calcium balance.[111] If the AN is prolonged, calcium deficiency alone can contribute to decreased BMD. Reversing the negative calcium balance in AN women is often difficult because of the inability to accurately identify the degree of calcium deficiency until after the refeeding process has begun.[111] Therefore, AN patients are recommended to consume at least 1,500 mg/day of $Ca^{2+}$ pending a more thorough assessment of their calcium status.

Clinical consensus favors obtaining the recommended $Ca^{2+}$ levels through diet. This recommendation is supported by the observation that intake of dairy products was more effective than calcium supplementation in improving bone density in $Ca^{2+}$-depleted rats.[111,112] However, compliance with diet and supplementation can be difficult for patients with AN in the inpatient setting, and even more so in the outpatient setting.[89,111] Therefore, supplementation is often required to reach the recommended intake levels.

Although the general recommendation is for patients to consume calcium through their diet and supplements, the role of calcium in AN treatment is uncertain, and there are competing reports on the effectiveness of calcium supplements on BMD in AN.[92] BMD can be partially restored with refeeding.[92] However, this is confounded by AN-related hormonal imbalances. For example, the BMD may not improve in many patients until menstrual function is restored despite adequate calcium intake.[92] More research is needed to determine whether calcium supplementation improves bone density and prevents additional bone loss in the presence of amenorrhea.

The effectiveness of calcium supplementation in AN women has been recently compared to bisphosphonate treatment. A study by Nakahara and colleagues (2006) compared etidronate to 600 mg calcium and vitamin D in 41 women with AN and found that either treatment was an effective means of increasing tibial BMD compared to placebo.[91] Although these results suggest that either bisphosphonates or calcium and vitamin D may be effective in treating AN-related osteopenia in the short term, long-term studies are still needed.

Another significant nutritional deficiency affecting the bone mass of AN patients is vitamin K deficiency. The low osteocalcin levels in anorexia are further confounded by simultaneous vitamin K deficiency, which results in loss of $\gamma$-carboxylation of osteocalcin (Fig. 6). The loss of $\gamma$-carboxylation decreases the affinity of osteocalcin for bone, and ultimately increases its function as a hormone on

pancreatic and adipose tissue, where it induces the expression of insulin and adiponectin, respectively (Fig. 6).[113] The use of menatetrenone (Vitamin K2) to increase the $\gamma$-carboxylation of osteocalcin has been shown in AN patients at a dose of 45 mg/day to slow decreases in BMD and increase markers of bone formation whereas decreasing markers of bone resorption.[114]

### Modulation of osteoclast activity by hormone replacement therapy

The data regarding the use of hormone replacement therapy has had varied results: scientific evidence has demonstrated partial correction of low bone mass, whereas clinical studies have demonstrated mixed results. Scientifically, the evidence suggests that hormone replacement should decrease osteoclastic activity in settings of alterations in CNS leptin/serotonin signaling.[99] The bone phenotype seen with leptin/serotonin alterations is not due to defects in bone resorption *per se*; however, the hypogonadism and resulting ostoclastic changes can be partially corrected through sex steroid normalization. For example, estradiol treatment of *ob/ob* leptin-deficient mice decreases the number of osteoclasts thereby increasing bone mass.[99]

Clinical evidence on sex steroid use has been mixed. AN patients that meet all the criteria for AN except for amenorrhea still have decreased lumbar and femoral BMD; this suggests that normal estrogen levels alone cannot protect bone mass in AN.[114] These findings are mirrored clinically: high-dose oral contraceptives are only protective of bone mass if they are associated with nutritional rehabilitation.[114] In general, most studies show that severely underweight patients are those who derive the greatest benefit from estrogen supplementation.[114]

### Conclusion

The pathology of bone loss in patients suffering from AN is a complex process consisting of multiple interconnected regulatory steps. With the realization that AN patients starve themselves of carbohydrate to reduce their serotonin levels, we have come to appreciate the effects of these changes on neural pathways and endocrine regulation. Leptin, through decreasing serotonin levels, acts to modulate sympathetic tone of the body—a process that directly impacts bone metabolism. The $\beta$-adrenergic system, acting downstream of leptin and serotonin, is likely a major determinant of bone mass in AN patients. However, further research is needed to confirm this hypothesis.

Despite major advances in the field, treatment for AN remains a challenge because of the lack of understanding of the underlying neuropsychiatric changes. Current therapies such as nutritional supplementation are greatly limited by the lack of compliance of patients suffering from AN. Although hormone replacement therapy and vitamin K administration can partially restore bone mass, these interventions fail to robustly improve BMD in the absence of adequate nutrition. Through better understanding of the impact of altered serotonin release, novel targets for treatment can be envisioned. Further characterization of these targets should reveal the potential for therapeutics that can reverse declines in bone mass and restore normal nutrition to patients suffering from AN.

## Acknowledgments

K.K.K. acknowledges support by the Public Health Service award T32 GM07347 from the National Institute of General Medical Studies for the Vanderbilt Medical Scientist Training Program. J.I. acknowledges prior research support from the American Federation of Aging Research.

## Conflicts of interest

The authors declare no conflicts of interest.

## References

1. Attia, E. 2010. Anorexia nervosa: current status and future directions. *Annu. Rev. Med.* **61:** 425–435.
2. Abell, T.L. *et al.* 1987. Gastric electromechanical and neurohormonal function in anorexia nervosa. *Gastroenterology* **93:** 958–965.
3. Gross, K.B., K.M. Brough & P.M. Randolph. 1986. Eating disorders: anorexia and bulimia nervosas. *ASDC J. Dent. Child.* **53:** 378–381.
4. Kaye, W.H., J.L. Fudge & M. Paulus. 2009. New insights into symptoms and neurocircuit function of anorexia nervosa. *Nat. Rev. Neurosci.* **10:** 573–584.
5. Jimerson, D. & B. Wolfe. 2006. Psychobiology of eating disorders in *Annual Review of Eating Disorders, Part 2*. Wonderlich, S., J., Mitchell, M. De Zwann, & H. Steiger, eds. 1–15 Radcliffe Publishing Ltd. Abingdon, UK.
6. Phillips, M.L. *et al.* 2003. Neurobiology of emotion perception I: the neural basis of normal emotion perception. *Biol. Psychiatry* **54:** 504–514.
7. Phillips, M.L. *et al.* 2003. Neurobiology of emotion perception II: implications for major psychiatric disorders. *Biol. Psychiatry* **54:** 515–528.
8. Gordon, I. *et al.* 1997. Childhood-onset anorexia nervosa: towards identifying a biological substrate. *Int. J. Eat. Disord.* **22:** 159–165.
9. Råstam, M. *et al.* 2001. Regional cerebral blood flow in weight-restored anorexia nervosa: a preliminary study. *Dev. Med. Child Neurol.* **43:** 239–242.
10. Uher, R. *et al.* 2003. Recovery and chronicity in anorexia nervosa: brain activity associated with differential outcomes. *Biol. Psychiatry* **54:** 934–942.
11. Havel, P.J. 2000. Role of adipose tissue in body-weight regulation: mechanisms regulating leptin production and energy balance. *Proc. Nutr. Soc.* **59:** 359–371.
12. Haluzikova, D. *et al.* 2009. Serum concentrations of adipocyte fatty acid binding protein in patients with anorexia nervosa. *Physiol. Res.* **58:** 577–581.
13. Margetic, S. *et al.* 2002. Leptin: a review of its peripheral actions and interactions. *Int. J. Obes. Relat. Metab. Disord.* **26:** 1407–1433.
14. Boden, G. *et al.* 1996. Effect of fasting on serum leptin in normal human subjects. *J. Clin. Endocrinol. Metab.* **81:** 3419–3423.
15. Considine, R. V. *et al.* 1996. Serum immunoreactive-leptin concentrations in normal-weight and obese humans. *N. Engl. J. Med.* **334:** 292–295.
16. Oswal, A. & G. Yeo. Leptin and the control of body weight: a review of its diverse central targets, signaling mechanisms, and role in the pathogenesis of obesity. *Obesity (Silver Spring)* **18:** 221–229.
17. Lee, G. H. *et al.* 1996. Abnormal splicing of the leptin receptor in diabetic mice. *Nature* **379:** 632–635.
18. Björbaek, C. *et al.* 1998. Expression of leptin receptor isoforms in rat brain microvessels. *Endocrinology* **139:** 3485–3491.
19. Hileman, S.M. *et al.* 2002. Characterizaton of short isoforms of the leptin receptor in rat cerebral microvessels and of brain uptake of leptin in mouse models of obesity. *Endocrinology* **143:** 775–783.
20. Tartaglia, L.A. 1997. The leptin receptor. *J. Biol. Chem.* **272:** 6093–6096.
21. Montague, C.T. *et al.* 1997. Congenital leptin deficiency is associated with severe early-onset obesity in humans. *Nature* **387:** 903–908.
22. Caro, J.F. *et al.* 1996. Decreased cerebrospinal-fluid/serum leptin ratio in obesity: a possible mechanism for leptin resistance. *Lancet* **348:** 159–161.
23. Considine, R.V. *et al.* 1995. Evidence against either a premature stop codon or the absence of obese gene mRNA in human obesity. *J. Clin. Invest.* **95:** 2986–2988.
24. Hassink, S.G. *et al.* 1996. Serum leptin in children with obesity: relationship to gender and development. *Pediatrics* **98:** 201–203.
25. Heymsfield, S.B. *et al.* 1999. Recombinant leptin for weight loss in obese and lean adults: a randomized, controlled, dose-escalation trial. *JAMA* **282:** 1568–1575.
26. Ghilardi, N. & R.C. Skoda. 1997. The leptin receptor activates janus kinase 2 and signals for proliferation in a factor-dependent cell line. *Mol. Endocrinol.* **11:** 393–399.

27. Banks, A.S. *et al.* 2000. Activation of downstream signals by the long form of the leptin receptor. *J. Biol. Chem.* **275:** 14563–14572.

28. Xu, A.W. *et al.* 2007. Inactivation of signal transducer and activator of transcription 3 in proopiomelanocortin (Pomc) neurons causes decreased pomc expression, mild obesity, and defects in compensatory refeeding. *Endocrinology* **148:** 72–80.

29. Morris, D. L. & L. Rui. 2009. Recent advances in understanding leptin signaling and leptin resistance. *Am. J. Physiol. Endocrinol. Metab.* **297:** E1247–E1259.

30. Chen, H. *et al.* 1996. Evidence that the diabetes gene encodes the leptin receptor: identification of a mutation in the leptin receptor gene in db/db mice. *Cell* **84:** 491–495.

31. Mercer, J.G. *et al.* 1996. Localization of leptin receptor mRNA and the long form splice variant (Ob-Rb) in mouse hypothalamus and adjacent brain regions by in situ hybridization. *FEBS Lett.* **387:** 113–116.

32. Thornton, J.E. *et al.* 1997. Regulation of hypothalamic proopiomelanocortin mRNA by leptin in ob/ob mice. *Endocrinology* **138:** 5063–5066.

33. Schwartz, M.W. *et al.* 1996. Identification of targets of leptin action in rat hypothalamus. *J. Clin. Invest.* **98:** 1101–1106.

34. Cowley, M.A. *et al.* 2001. Leptin activates anorexigenic POMC neurons through a neural network in the arcuate nucleus. *Nature* **411:** 480–484.

35. Figlewicz, D.P. & S.C. Benoit. 2009. Insulin, leptin, and food reward: update 2008. *Am. J. Physiol. Regul. Integr. Comp. Physiol.* **296:** R9–R19.

36. DiLeone, R.J., D. Georgescu & E.J. Nestler. 2003. Lateral hypothalamic neuropeptides in reward and drug addiction. *Life Sci.* **73:** 759–768.

37. Harris, G.C., M. Wimmer & G. Aston-Jones. 2005. A role for lateral hypothalamic orexin neurons in reward seeking. *Nature* **437:** 556–559.

38. Huszar, D. *et al.* 1997. Targeted disruption of the melanocortin-4 receptor results in obesity in mice. *Cell* **88:** 131–141.

39. Chen, A.S. *et al.* 2000. Inactivation of the mouse melanocortin-3 receptor results in increased fat mass and reduced lean body mass. *Nat. Genet.* **26:** 97–102.

40. Balthasar, N. *et al.* 2005. Divergence of melanocortin pathways in the control of food intake and energy expenditure. *Cell* **123:** 493–505.

41. Coll, A.P., I.S. Farooqi & S.O'Rahilly. 2007. The hormonal control of food intake. *Cell* **129:** 251–262.

42. Coll, A.P. *et al.* 2008. SnapShot: the hormonal control of food intake. *Cell* **135:** e1–e2.

43. Gropp, E. *et al.* 2005. Agouti-related peptide-expressing neurons are mandatory for feeding. *Nat. Neurosci.* **8:** 1289–1291.

44. Pinto, S. *et al.* 2004. Rapid rewiring of arcuate nucleus feeding circuits by leptin. *Science* **304:** 110–115.

45. Sternson, S.M., G.M. Shepherd & J.M. Friedman. 2005. Topographic mapping of VMH –> arcuate nucleus microcircuits and their reorganization by fasting. *Nat. Neurosci.* **8:** 1356–1363.

46. Tung, Y.C. *et al.* 2008. Novel leptin-regulated genes revealed by transcriptional profiling of the hypothalamic paraventricular nucleus. *J. Neurosci.* **28:** 12419–12426.

47. Holder, J.L. Jr., N.F. Butte & A.R. Zinn. 2000. Profound obesity associated with a balanced translocation that disrupts the SIM1 gene. *Hum. Mol. Genet.* **9:** 101–108.

48. Michaud, J.L. *et al.* 2001. Sim1 haploinsufficiency causes hyperphagia, obesity and reduction of the paraventricular nucleus of the hypothalamus. *Hum. Mol. Genet.* **10:** 1465–1473.

49. Michaud, J.L. *et al.* 1998. Development of neuroendocrine lineages requires the bHLH-PAS transcription factor SIM1. *Genes Dev.* **12:** 3264–3275.

50. Kaye, W.H. *et al.* 1987. Elevated cerebrospinal fluid levels of immunoreactive corticotropin-releasing hormone in anorexia nervosa: relation to state of nutrition, adrenal function, and intensity of depression. *J. Clin. Endocrinol. Metab.* **64:** 203–208.

51. Lesem, M.D. *et al.* 1994. Cerebrospinal fluid TRH immunoreactivity in anorexia nervosa. *Biol. Psychiatry* **35:** 48–53.

52. Tran, P.V. *et al.* 2006. Diminished hypothalamic bdnf expression and impaired VMH function are associated with reduced SF-1 gene dosage. *J. Comp. Neurol.* **498:** 637–648.

53. Cao, L. *et al.* 2010. Environmental and genetic activation of a brain-adipocyte BDNF/leptin axis causes cancer remission and inhibition. *Cell* **142:** 52–64.

54. Unger, T.J. *et al.* 2007. Selective deletion of Bdnf in the ventromedial and dorsomedial hypothalamus of adult mice results in hyperphagic behavior and obesity. *J. Neurosci.* **27:** 14265–14274.

55. Yeo, G.S *et al.* 2004. A de novo mutation affecting human TrkB associated with severe obesity and developmental delay. *Nat. Neurosci.* **7:** 1187–1189.

56. Fairbanks, L. *et al.* 2001. Social impulsivity inversely associated with CSF 5-HIAA and fluoxetine exposure in vermet monkeys. *Neuropsychopharmacology* **24:** 370–378.

57. Lesch, K. & U. Merschdorf. 2000. Impulsivity, aggression, and serotonin: a molecular psychobiological perspective. *Behav. Sci. Law* **185:** 581–604.

58. Mann, J.J. 1999. Role of the serotonergic system in the pathogenesis of major depression and suicidal behavior. *Neuropsychopharmacology* **21:** S99–S105.

59. Simansky, K.J. 1996. Serotonergic control of the organization of feeding and satiety. *Behav. Brain Res.* **73:** 37–42.

60. Soubrie, P. 1986. Reconciling the role of central serotonin neurons in human and animal behavior. *Behav. Brain Sci.* **9:** 319–364.

61. Fernstrom, J.D. & R.J. Wurtman. 1972. Brain serotonin content: physiological regulation by plasma neutral amino acids. *Science* **178:** 414–416.

62. Kaye, W.H. 2003. Anxiolytic effects of acute tryptophan depletion in anorexia nervosa. *Int. J. Eat. Disord.* **33:** 257–267.

63. Crisp, A.H. 1978. Disturbances of neurotransmitter metabolism in anorexia nervosa. *Proc. Nutr. Soc.* **37:** 201–209.

64. Stanley, M., L. Traskman-Bendz & K. Dorovini-Zis. 1985. Correlations between aminergic metabolites simultaneously obtained from human CSF and brain. *Life Sci.* **37:** 1279–1286.

65. Audenaert, K. 2003. Decreased 5-HT2a receptor binding in patients with anorexia nervosa. *J. Nucl. Med.* **44:** 163–169.

66. Bailer, U.F. 2004. Altered 5-HT2A receptor binding after recovery from bulimia-type anorexia nervosa: relationships to harm avoidance and drive for thinness. *Neuropsychopharmacology.* **29:** 1143–1155.

67. Bailer, U.F. *et al.* 2007. Exaggerated 5-HT1A but normal 5-HT2A receptor activity in individuals ill with anorexia nervosa. *Biol. Psychiatry.* **61:** 1090–1099.

68. Bailer, U.F. *et al.* 2005. Altered brain serotonin 5-HT1A receptor binding after recovery from anorexia nervosa measured by positron emission tomography and [carbonyl11C]WAY-100635. *Arch. Gen. Psychiatry.* **62:** 1032–1041.

69. Frank, G.K. 2002. Reduced 5-HT2A receptor binding after recovery from anorexia nervosa. *Biol. Psychiatry.* **52:** 896–906.

70. Galusca, B. *et al.* 2008. Organic background of restrictive-type anorexia nervosa suggested by increased serotonin 1A receptor binding in right frontotemporal cortex of both lean and recovered patients: [18F]MPPF PET scan study. *Biol. Psychiatry.* **64:** 1009–1013.

71. Tiihonen, J. *et al.* 2004. Brain serotonin 1A receptor binding in bulimia nervosa. *Biol. Psychiatry.* **55:** 871–873.

72. Cloninger, C.R. *et al.* 1994. *The Temperament and Character Inventory (TCI): A Guide To Its Development and Use.* St. Louis, MO: Center for Psychobiology of Personality, Washington University.

73. Yadav, V.K. *et al.* 2009. A serotonin-dependent mechanism explains the leptin regulation of bone mass, appetite, and energy expenditure. *Cell* **138:** 976–989.

74. Frank, G. 2005. Increased dopamine D2/D3 receptor binding after recovery from anorexia nervosa measured by positron emission tomography and [11C]raclopride. *Biol. Psychiatry* **58:** 908–912.

75. Bergen, A. 2005. Association of multiple DRD2 polymorphisms with anorexia nervosa. *Neuropsychopharmacology* **30:** 1703–1710.

76. Kaye, W.H., G.K. Frank & C. McConaha. 1999. Altered dopamine activity after recovery from restricting-type anorexia nervosa. *Neuropsychopharmacology* **21:** 503–506.

77. DeDeurwaerdere, P. *et al.* 2004. Constitutive activity of the serotonin2C receptor inhibits in vivo dopamine release in the rat striatum and nucleus accumbens. *J. Neurosci.* **24:** 3235–3241.

78. Di Matteo, V. *et al.* 2000. Biochemical and electrophysiological evidence that RO 60–0175 inhibits mesolimbic dopaminergic function through serotonin2C receptors. *Brain Res.* **865:** 85–90.

79. Rask-Andersen, M. *et al.* Molecular mechanisms underlying anorexia nervosa: focus on human gene association studies and systems controlling food intake. *Brain Res. Rev.* **62:** 147–164.

80. Leibowitz, S.F., P. Roossin & M. Rosenn. 1984. Chronic norepinephrine injection into the hypothalamic paraventricular nucleus produces hyperphagia and increased body weight in the rat. *Pharmacol. Biochem. Behav.* **21:** 801–808.

81. Heal, D.J. *et al.* 1998. Sibutramine: a novel anti-obesity drug. A review of the pharmacological evidence to differentiate it from d-amphetamine and d-fenfluramine. *Int. J. Obes. Relat. Metab. Disord.* **22**(Suppl. 1): S18–S28; discussion S29.

82. Kaye, W.H. *et al.* 1985. Disturbances of norepinephrine metabolism and alpha-2 adrenergic receptor activity in anorexia nervosa: relationship to nutritional state. *Psychopharmacol Bull.* **21:** 419–423.

83. Kaye, W.H. *et al.* 1985. Altered norepinephrine metabolism following long-term weight recovery in patients with anorexia nervosa. *Psychiatry Res.* **14:** 333–342.

84. Kaye, W.H. *et al.* 1984. Abnormalities in CNS monoamine metabolism in anorexia nervosa. *Arch. Gen. Psychiatry.* **41:** 350–355.

85. Bartak, V. *et al.* 2004. Basal and exercise-induced sympathetic nervous activity and lipolysis in adipose tissue of patients with anorexia nervosa. *Eur. J. Clin. Invest.* **34:** 371–377.

86. Nedvidkova, J. *et al.* 2004. Increased subcutaneous abdominal tissue norepinephrine levels in patients with anorexia nervosa: an in vivo microdialysis study. *Physiol. Res.* **53:** 409–413.

87. Chatelain, P. *et al.* 1992. Prevention of calcium overload and down-regulation of calcium channels in rat heart by SR 33557, a novel calcium entry blocker. *Cardioscience* **3:** 117–123.

88. Nakagawa, K. *et al.* 1987. Beta-adrenergic receptor in heart of starved rats. *Endocrinol. Jpn.* **34:** 381–385.

89. Brooks, E.R., P.M. Howat & D.S. Cavalier. 1999. Calcium supplementation and exercise increase appendicular bone density in anorexia: a case study. *J. Am. Diet. Assoc.* **99:** 591–593.

90. Brotman, A.W. & T.A. Stern. 1985. Osteoporosis and pathologic fractures in anorexia nervosa. *Am. J. Psychiatry.* **142:** 495–496.

91. Nakahara, T. *et al.* 2006. The effects of bone therapy on tibial bone loss in young women with anorexia nervosa. *Int. J. Eat. Disord.* **39:** 20–26.

92. Salisbury, J.J. & J.E. Mitchell. 1991. Bone mineral density and anorexia nervosa in women. *Am. J. Psychiatry.* **148:** 768–774.

93. Oswiecimska, J. *et al.* 2007. Skeletal status and laboratory investigations in adolescent girls with anorexia nervosa. *Bone* **41:** 103–110.

94. Iqbal, J., L. Sun & M. Zaidi. Commentary-FSH and bone 2010: evolving evidence. *Eur. J. Endocrinol.* **163:** 173–176.

95. Milos, G. *et al.* 2005. Cortical and trabecular bone density and structure in anorexia nervosa. *Osteoporos. Int.* **16:** 783–790.

96. Angeli, A. *et al.* 1972. Behavior of the circadian rhythm of adrenal cortex secretion in anorexia nervosa in relation to the duration of amenorrhea. *Folia Endocrinol.* **25:** 153–160.

97. Bachrach, L. K. *et al.* 1990. Decreased bone density in adolescent girls with anorexia nervosa. *Pediatrics* **86:** 440–447.

98. Milos, G. *et al.* 2007. Does weight gain induce cortical and trabecular bone regain in anorexia nervosa? A two-year prospective study. *Bone* **41:** 869–874.

99. Ducy, P. *et al.* 2000. Leptin inhibits bone formation through a hypothalamic relay: a central control of bone mass. *Cell* **100:** 197–207.

100. Takeda, S. *et al.* 2002. Leptin regulates bone formation via the sympathetic nervous system. *Cell* **111:** 305–317.

101. Rosen, C.J. 2009. Leptin's RIGHT turn to the brain stem. *Cell Metab.* **10:** 243–244.

102. Iqbal, J. *et al.* 2006. Follicle-stimulating hormone stimulates TNF production from immune cells to enhance osteoblast and osteoclast formation. *Proc. Natl. Acad. Sci. USA* **103:** 14925–14930.

103. Sun, L. *et al.* 2006. FSH directly regulates bone mass. *Cell* **125:** 247–260.

104. Swerdloff, R.S., R.A. Batt & G.A. Bray. 1976. Reproductive hormonal function in the genetically obese (ob/ob) mouse. *Endocrinology* **98:** 1359–1364.

105. Devleta, B., B. Adem & S. Senada. 2004. Hypergonadotropic amenorrhea and bone density: new approach to an old problem. *J. Bone Miner. Metab.* **22:** 360–364.

106. Zainabadi, K. 2009. SirT1 regulates bone mass in vivo through regulation of osteoblast andosteoclast differentiation. PhD, MIT. Boston.

107. Cakir, I. *et al.* 2009. Hypothalamic Sirt1 regulates food intake in a rodent model system. *PLoS One.* **4:** e8322, 1–12.

108. Ramadori, G. *et al.* SIRT1 deacetylase in POMC neurons is required for homeostatic defenses against diet-induced obesity. *Cell Metab.* **12:** 78–87.

109. Kalra, S.P., M.G. Dube & U.T. Iwaniec. 2009. Leptin increases osteoblast-specific osteocalcin release through a hypothalamic relay. *Peptides.* **30:** 967–973.

110. DeMambro, V.E. *et al.* 2008. Gender-specific changes in bone turnover and skeletal architecture in igfbp-2-null mice. *Endocrinology* **149:** 2051–2061.

111. Setnick, J. Micronutrient deficiencies and supplementation in anorexia and bulimia nervosa: a review of literature. *Nutr. Clin. Pract.* **25:** 137–142.

112. Weaver, K., J. Wuest & D. Ciliska. 2005. Understanding women's journey of recovering from anorexia nervosa. *Qual. Health Res.* **15:** 188–206.

113. Lee, N.K. *et al.* 2007. Endocrine regulation of energy metabolism by the skeleton. *Cell* **130:** 456–469.

114. Bruni, V., M.F. Filicetti & V. Pontello. 2006. Open issues in anorexia nervosa: prevention and therapy of bone loss. *Ann. N.Y. Acad. Sci.* **1092:** 91–102.

Ann. N.Y. Acad. Sci. ISSN 0077-8923

# ANNALS OF THE NEW YORK ACADEMY OF SCIENCES
Issue: *Molecular and Integrative Physiology of the Musculoskeletal System*

# Bone and muscle loss after spinal cord injury: organ Interactions

Weiping Qin,[1,2] William A. Bauman,[1,2,3] and Christopher Cardozo[1,2,3]

[1]Center of Excellence for the Medical Consequences of Spinal Cord Injury, James J. Peters Veterans Affairs Medical Center, Bronx, New York. [2]Department of Medicine and [3]Department of Rehabilitation Medicine, Mount Sinai School of Medicine, New York, New York

Address for correspondence: Weiping Qin, M.D., Ph.D., Center of Excellence for the Medical Consequences of Spinal Cord Injury, James J. Peters VA Medical Center, 130 West Kingsbridge Road, Bronx, NY 10468. Weiping.qin@mssm.edu

Spinal cord injury (SCI) results in paralysis and marked loss of skeletal muscle and bone below the level of injury. Modest muscle activity prevents atrophy, whereas much larger—and as yet poorly defined—bone loading seems necessary to prevent bone loss. Once established, bone loss may be irreversible. SCI is associated with reductions in growth hormone, IGF-1, and testosterone, deficiencies likely to exacerbate further loss of muscle and bone. Reduced muscle mass and inactivity are assumed to be contributors to the high prevalence of insulin resistance and diabetes in this population. Alterations in muscle gene expression after SCI share common features with other muscle loss states, but even so, show distinct profiles, possibly reflecting influences of neuromuscular activity due to spasticity. Changes in bone cells and markers after SCI have similarities with other conditions of unloading, although after SCI these changes are much more dramatic, perhaps reflecting the much greater magnitude of unloading. Adiposity and marrow fat are increased after SCI with intriguing, though poorly understood, implications for the function of skeletal muscle and bone cells.

Keywords: spinal cord injury; bone loss; muscle atrophy; body composition; testosterone; estrogen; insulin-like growth factor 1; methylprednisolone

## Introduction

Estimates from the National Spinal Cord Injury Database (https://www.nscisc.uab.edu/public_content/annual_stat_report.aspx) indicate that 220,000–300,000 individuals with spinal cord injury (SCI) reside in the United States, and that about 12,000 new spinal cord injuries occur annually. Approximately 56% of the cases have tetraplegia from injury to the cervical cord with the remainder having paraplegia from injury to lower vertebral sites.[1] A significant proportion of persons with SCI have complete loss of motor and sensory function below the level of SCI (40–42%).[2] The most frequent cause of injury is traffic accidents (43–50%), followed by falls, sports accidents, and acts of violence. Males are about four times more likely to have SCI than are females. Injuries most commonly occur for both genders between the ages of 15 and 24 years. The fiscal burden of individuals

with SCI has been estimated to be greater than $14.5 billion annually in direct medical costs and disability support.[3]

Disruption of the long tracts of the spinal cord results in varying degrees of paralysis and loss of sensation below the level of the lesion. Importantly, a state of extreme immobilization may ensue with dramatic, and virtually unparalleled, loss of bone and skeletal muscle.[4–6] These alterations are accompanied by the development of the numerous secondary medical consequences that include dysfunction of the pituitary–hypothalamic axis, increased adiposity, insulin resistance, spasticity of paralyzed regions, and autonomic dysfunction.[7–12] When the injury occurs above the level of the sixth thoracic vertebra, preganglionic sympathetic neurons are severed, resulting in loss of sympathetic outflow, which is more complete as the level of injury ascends through the thoracic vertebrae, and is associated with complete sympathectomy for injuries

doi: 10.1111/j.1749-6632.2010.05806.x

above the first thoracic vertebra. Thus, at higher vertebral levels, SCI disrupts sympathetic innervation to the heart, lung, vasculature, skeletal muscle, bone, and adrenal glands, and to other tissues.[13]

Preservation and/or restoration of sufficient muscular function and skeletal integrity to support weight bearing and ambulation are central challenges that must be overcome before interventions that repair the injured spinal cord, or use interim technologies, such as functional electrical stimulation (FES) or activation of lumbar pattern generators, can be effectively and safely implemented. In this review, we will discuss recent advances in our understanding of physiological, biochemical, and molecular alterations in skeletal muscle and bone after SCI, and we will examine the interactions between muscle, bone, and other tissues and organs as they relate to the loss of muscle and bone.

## Muscle and bone after SCI

### Skeletal muscle

The characteristics of the alterations in skeletal muscle mass, physiology, and fiber type have been recently reviewed in detail[4,5] and, therefore, will only be summarized here. Following SCI, skeletal muscle atrophies by approximately 30–60%, depending on muscle type and completeness of the lesion, associated with reductions in maximal force generation and endurance. Atrophy is associated with a switch from type I slow oxidative to type II fast glycolytic fibers, a process that begins within several months after SCI and evolves over several years. After SCI, muscle demonstrates fundamental changes in physiological properties that include markedly increased fatigability and altered rates of contraction and relaxation. Intriguingly, in untrained individuals with SCI, exercise of paralyzed muscles using FES results in greater tissue damage compared to similar FES exercise performed in controls, when assessed by MRI;[14] why this greater predisposition to muscle injury occurs after surface electrical stimulation of sublesional muscle is unknown. A possible molecular mechanism is suggested by findings that mice lacking peroxisome proliferator receptor gamma coactivator-1α (PGC-1α), a critical regulator of oxidative metabolism[15,16] known to be downregulated in disuse atrophy,[17] display greater activity-induced muscle injury than controls.[18]

Changes in gene expression after SCI have been characterized using gene expression profiling in hu-

mans within the first week after injury, as well as in rat models. In both animals and humans, changes in gene expression after SCI share several common features with other forms of disuse atrophy, including elevations of two muscle-restricted E3 ubiquitin ligases, muscle atrophy F-box (MAFbx; also called atrogin-1), and muscle ring finger 1 (MuRF1).[17,19,20] Loss of either of these E3 ubiquitin ligases has been shown to slow denervation atrophy, and loss of MuRF1 prevents glucocorticoid-induced destruction of myosin heavy chain (MyHC).[21,22]

In the spinal isolation model, a variant of SCI in which reflex arcs responsible for spasticity are disrupted by dorsal rhizotomy, many of the alterations in gene expression observed were similar to those found after nerve transection, as well as to those seen in fasting, uremia, diabetes, or cancer cachexia.[17] These gene changes include increased expression of MAFbx and MuRF1, as well as that of the transcription factor FOXO1, a key factor in upregulation of MAFbx, MuRF1, and of 4EBP1, an inhibitor of translation initiation.[17] Reduced expression of the transcriptional coregulator PGC-1α, which protects against muscle atrophy,[15,16,23] was also found in both spinal isolation and denervation.[17]

Differences exist between models of neurological injury, including those of SCI, spinal isolation, and nerve transection (e.g., denervation). Levels of Runx1, a transcription factor expressed in embryonic muscle that confers protection against denervation atrophy,[24] and the myogenic transcription factor myogenin, return to baseline within 7–14 days after SCI but remain dramatically elevated after denervation.[19] Levels of MAFbx and MuRF1 return to baseline values within two weeks after SCI, but more gradually return to baseline levels after nerve transection.[19,25]

The marked decline in expression of MAFbx, MuRF1, myogenin, and Runx1 within two weeks after SCI correlates to the emergence of spasticity,[19] notwithstanding important effects of an intact motor neuron and motor end plate. This point is illustrated by findings from microarray studies that found differences in gene expression profiles between spinal isolation and nerve transection with genes involved in proteolysis, glycolysis, ATP synthesis, and extracellular matrix declining only in spinal isolation.[17] These differences may reflect the small residual electrical activity of the motor neuron in the spinal isolation model, the trophic influence of the nerve, or consequences of the organization of the motor endplate.

Consistent with the view that immobilization is the primary stimulus to muscle atrophy after SCI, many of the changes in muscle mass, gene expression, and physiological properties have been shown to be prevented, or reversed, by FES. In persons with a SCI consistent with the American Spinal Injury Association (ASIA) Impairment Scale A classification (e.g., complete motor and sensory loss), twice-weekly resistance training using FES significantly increased muscle size and endurance, with benefits to endurance reaching a plateau by 12 weeks.[26] Exercise using FES prevented subsequent atrophy when begun within 14–15 weeks after acute SCI, as assessed by DXA (dual energy X-ray absorptiometry) scanning.[27] Similarly, studies in rats with spinal isolation demonstrated that several brief periods of contractile activity each day, for a total of 30 min/day, blocked loss of muscle mass and upregulation of MAFbx and MuRF1.[28]

A fundamental question that remains incompletely answered is how electromechanical coupling switches off atrophy signaling programs and promotes muscle hypertrophy. Potential mechanisms include:

1. upregulation of trophic and growth factors such as insulin-like growth factor (IGF)-1;[29]
2. coupling of the sarcomere and nucleus through serum response factor (SRF) mediated by activation of titin kinase by mechanically-induced conformational changes,[30] as well as multiple changes in activity of other kinases;[31] and
3. possible roles for calcium fluxes and mechanical loading sensed through integrins.

In healthy muscle, there is a clearly delineated role for calcium-dependent signaling in specification of muscle fiber type. Such signaling is mediated largely through the protein phosphatase, calcineurin, and its substrate, nuclear factor of activated T cells (NFAT; a transcription factor). NFAT acts in concert with PGC-1α to promote formation of oxidative fibers.[32–34] The role of calcium-dependent signaling through calcineurin in maintaining skeletal muscle mass remains controversial.[32,35–38] The possibility that calcineurin may play essential roles in determining muscle mass in

states of atrophy is suggested by an association in the chronically denervated rat muscle between reduced expression of the calcineurin inhibitor RCAN2 and delayed muscle atrophy (Qin *et al.*, in press). Of potential relevance to understanding the role of calcineurin signaling in maintaining muscle mass under conditions predisposing to muscle atrophy, cardiac and skeletal muscle share many mechanisms responsible for atrophy and hypertrophy, and altered calcineurin activity has been suggested to be an important contributor to the hypertrophy of cardiac myocytes in pathological conditions.[39]

Much has been learned about the molecular mechanisms that promote catabolism of muscle proteins during muscle atrophy. Much less is understood about the molecular basis for phenotype changes in muscle, such as the shift toward fast, glycolytic fibers. One possible mechanism for muscle fiber phenotypic change is reduced expression of PGC-1α, as reported in the spinal isolation model.[17] Because increased PGC-1α transcriptional activity in skeletal muscle is increased by exercise,[40] it might be inferred that the absence of muscle activity furthers any negative effects of decreased levels of PGC-1α on transcriptional activity in skeletal muscle after the SCI-associated immobilization.

Several NFAT isoforms are found in skeletal muscle, but it has been unclear why multiple isoforms are needed. A recent examination of the roles of the NFAT isoforms c1–c4 in expression of different MyHC forms indicated that these four isoforms have distinct functions in transcriptional regulation of skeletal muscle genes.[41] Only NFATc1 activated slow oxidative MyHC genes in response to low-frequency electrical stimulation, whereas NFATc2 and c3 were activated by higher frequency stimulation and led to the expression of fast oxidative and fast glycolytic MyHC genes.[41] An intriguing and poorly understood finding from this work was that NFATc4 is constitutively present in the nucleus of healthy skeletal muscle, inferring a more fundamental role in determining skeletal muscle gene expression. The extent to which calcineurin/NFAT signaling is dysregulated in SCI, or the roles for PGC-1α and NFAT in fiber type switching, remain to be addressed. Whatever the molecular basis, these alterations do not appear permanent. One year of FES-exercise partially reversed the fiber-type switching observed after SCI from slow to fast glycolytic fibers,[42] although fiber-type reversion has not been observed in other reports using FES-exercise of shorter duration [reviewed in Ref. 4].

## Bone

Bone loss after SCI is relatively unique in its magnitude, rapidity, and localization to the sublesional regions. The reader is referred to the recent reviews[4,43] on this topic for a detailed examination of the literature. In neurologically complete motor SCI, bone loss proceeds at a rate of 1% per week for the first 6–12 months,[44–46] a rate that is substantially greater than that observed during microgravity (0.25%/week),[47] prolonged bed rest (0.1%/week),[48] and after menopause in women who are not taking osteo-protective drugs and lose about 3–5% per year.[49] Rapid bone loss continues for at least 3–7 years after acute SCI, depending upon the bone region and parameter studied.[50] From a cross-sectional monozygotic twin study, discordant for chronic SCI, it has been suggested that bone loss in persons with SCI continues at an accelerated rate for decades.[51]

In a cross-sectional study of eight SCI patients, a change of 35.3% in bone mineral density (BMD) of the tibial trabecular bone was noted within the first 2 years after SCI, whereas there was only a 12.9% reduction in tibial cortical bone.[52] In a cross-sectional study of 31 patients with SCI for greater than 1 year, Dauty and colleagues[53] demonstrated a demineralization of −52% for the distal femur and −70% for the proximal tibia, in agreement with the findings of several other investigators.[50,54–56] In a cross-sectional study of men with motor complete SCI, bone loss in the epiphysis, which is predominantly trabecular bone, was exponential with time.[50] On average, loss of ephiphyseal bone was 50% in the femur and 60% in the tibia.[50] In the diaphyses or shaft, which is predominantly cortical bone, losses were 35% in the femur and 25% in the tibia, and involved erosion of the thickness of cortical bone by 0.25 mm/year over the initial 5–7 years after SCI.[50] Bone demineralization of the lumbar spine occurs after acute SCI, but the bone loss appears to be considerably less than that of the appendicular skeleton in the sublesional regional sites. Spontaneous pathological fractures occur in persons with chronic SCI, most often affecting the distal femur or proximal tibia. It has been reported that the risk of fractures is most closely related to the BMD of epiphyseal trabecular bone[57] and is directly correlated

with the duration of injury,[58] suggesting increased sublesional skeletal fragility with advancing age after SCI. The degree of neurological impairment has also been related to the risk of fractures.[59]

## Bone, muscle, and mechanical unloading

Bone responds to the mechanical loading of weight bearing, impacts such as heel strike, and deformation in tension, compression, and torsion from muscle contraction, by an increase in net bone formation. Immobilization removes or greatly attenuates these forms of bone loading and in all likelihood, is the primary cause of SCI-related bone loss. Neuronal, endocrine, and lifestyle factors provide additional, though relatively minor, influences that predispose to, or worsen, the rapidity of SCI-related bone loss. Consistent with this view, in an animal model of muscle and bone loss due to paralysis from transection of the sciatic nerve, protection against bone loss by clenbuterol depended upon the preservation of skeletal muscle mass and function.[60] This is supported by the finding that protection was lost when the triceps surae tendon was cut.[60] Similarly, after SCI, leg muscle mass loss was correlated with bone loss[61] and increased bone loading equivalent to 110–140% of body weight, using FES-induced soleus contraction, slowed bone loss after acute SCI.[62–64] An intriguing finding from this work, which employed FES techniques eliciting isolated soleus muscle contraction without cocontraction of other calf muscles, was that regions of preserved bone were limited primarily to posterior aspects of the tibia.[62] Ongoing studies are now examining the effect of activity-based rehabilitation performed in association with electrical stimulation on bone and muscle atrophy in subjects with subacute SCI in which electrical stimulation elicits cocontraction of flexor and extensor muscles of the legs. Preliminary findings suggest greater muscle mass and bone density in individuals receiving these interventions (Gail F. Forrest, personal communication, Kessler Foundation Research Center).

For unclear reasons, the technique of isolated soleus contraction did not appear to increase bone in individuals with chronic SCI.[65] However, one year of FES-induced cycling increased tibial bone by 10%.[66] Standing alone or locomotor training using body weight supported treadmill training have been insufficient to stimulate increases in bone.[4,67–69] While one might suppose that muscle activity is protective against bone loss, the degree of spasticity after SCI has correlated with BMD after injury in some studies but not others.[56,70,71] One important difference between these studies is sample size, which was largest (see Ref. 54) in a study suggesting a relationship between spasticity and bone mass,[71] but much smaller in those that did not find such a correlation (samples sizes were 24 and 18 for Refs 56 and 70, respectively). While muscle and bone are coupled, this coupling does not appear to be absolute; spasticity or interventions with fairly minimal stimuli preserve muscle,[72] while marked loading is required just for partial protection from bone loss.[62–64] Exertion using FES is effective in increasing muscle mass, but passive weight-bearing[4] and FES-exercise training[65,66] appear to be less effective in increasing bone mass during chronic SCI. A difference between standing and FES exercise studies is the cyclical loading of bone in the latter. Other considerations may include a need for exercise paradigms that recruit greater muscle mass and achieve higher loading that would be more representative of physiological loads or, potentially, correction of inappropriate sensing of load in bone or propagation of load-induced cellular signals.

Mechanisms by which mechanical loading of bone is sensed, or by which this initial signal is propagated to support bone remodeling, remain incompletely understood. The current state of knowledge on this topic has been reviewed elsewhere.[73,74] It is understood that osteocytes, compartmentalized within the matrix, sense and respond to changes in canalicular fluid flow arising from deformation of bone due to mechanical loads.[73,75] These bone cells communicate with each other through connexin-mediated gap junctions in filipodial extensions within the canalicular network. Mechanical stress, sensed by the osteocyte, stimulates bone formation and inhibits bone resorption, thereby increasing bone mass and strength and maintaining bone structural integrity.

The cellular and molecular events involve intracellular signal transmitters ($Ca^{2+}$, IP3, cAMP, and cGMP) and extracellular signal transmitters ($PGE_2$, $PGI_2$, IGF-1, IGF-2, and TGF-ß) to induce bone formation by osteoblasts, inhibition of bone resorption by osteoclasts, or a combination of the two.[76] Emerging evidence suggests that the Wnt/beta-catenin pathway in osteocytes may be triggered by crosstalk with the prostaglandin pathway in response to loading, which then leads to a decrease in expression of negative regulators of the pathway, such as sclerostin, the product of the SOST gene (in the past, also referred to as the sclerosteosis gene) and Dkk1.[77] Mechanical strain of mesenchymal precursor cells has been shown to determine the lineage commitment. In a cell culture system using mesenchymal precursors, cellular strain led to an osteoblastic phenotype through activation of RhoA/ROCK, whereas absence of stretch predisposed to an adipocyte phenotype.[78] Whether loading of precursor cells in marrow after SCI assists in determining their lineage is unknown.

## Bone cells and markers

Studies of calcium metabolism, blood markers of bone formation and breakdown, and properties of bone cells indicate that, following SCI, a state of hyper-resorption of bone ensues, associated with hypercalciuria. Increased levels of alkaline phosphatase have been reported during the first postinjury year.[79,80] Levels of osteocalcin, another marker of bone formation, were within the normal range for the first month after injury, with levels peaking above the normal range several months thereafter.[81] In a 24-week study after acute SCI, total alkaline phosphatase levels were again depressed immediately after injury but gradually rose to above the normal range by week 16, with osteocalcin levels also rising, albeit in the normal range.[80] Several bone formation markers were reported to be low early after SCI but then again increased, especially the amino-terminal propeptide of type III collagen.[82]

Markers of bone resorption found to be elevated after SCI include urinary free and total pyridinoline and deoxypyridinoline (DPD) cross-links, serum type 1 collagen C-telopeptide (CTX), and N-telopeptide (NTX).[83] After SCI, marked increases in these bone resorption markers have been reported to occur as early as two weeks, reaching peak values two to four months after injury,[43] and remaining elevated for at least six months postinjury.[80] These reports on metabolic bone markers suggest an initial acute depression of bone turnover with a subsequent, gradual increase in bone turnover, possibly initially coupled to the heightened resorption during the subacute phase of injury. A cross-sectional study has shown mildly elevated levels of DPD in 30% of

SCI individuals for over 10 years.[58] Another report found a positive correlation of CTX with increasing age.[84] Although conjecture, a heightened state of bone turnover in persons with chronic SCI may, in part, be a consequence of vitamin D deficiency and a state of mild chronic secondary hyperparathyroidism.

Less is known about the properties of bone cells after SCI. In a study of bone cells derived from bone marrow biopsy obtained above (sternum) or below (iliac crest) the level of lesion in subjects with paraplegia, osteoclast number was found to increase below the level of injury; media conditioned by marrow cells from below the injury promoted the formation of osteoclasts and contained increased levels of interleukin (IL)-6, tumor necrosis factor (TNF)-$\alpha$, and $PGE_2$.[85]

Studies of the properties of bone and bone cells after SCI have been reported in several animal models. These reports appear to recapitulate the changes in bone mass and structure observed in humans and they demonstrate alterations in biology and gene expression of bone cells consistent with the clinical picture of a state of accelerated bone resorption. Mechanical properties of bone were studied 24 weeks after complete SCI in male Sprague–Dawley rats;[86] reductions in relative bone strength of 63% were observed for the femoral and tibial shafts. The authors noted that these reductions in bone strength were similar in magnitude to BMD changes reported in humans with chronic paraplegia.[86] Ten days after contusion (e.g., incomplete) SCI in juvenile male rats, decreases in trabecular and cortical bone of 48% and 35%, respectively, were observed at the distal femoral metaphysis by micro-CT, and a 330% increase in the number of mature osteoclasts was observed at the growth plate.[87]

In a separate study, properties of bone-marrow cell-derived osteoblasts and osteoclasts from male rats three weeks after complete SCI were analyzed.[88] There were increases in osteoclast number, bone resorption in pit-forming assays, and expression of receptor activator for nuclear factor $\kappa$B ligand (RANKL) mRNA and protein by osteoblasts. The number of osteoblasts was unchanged. Gene expression for osteoprotegerin (OPG) was modestly reduced, while levels of alkaline phosphatase and osterix were unchanged. Paradoxically, tetracycline-labeling studies indicated that both bone formation and bone resorption were increased at three weeks.

In a subsequent time course study, bone formation rates declined to normal by six weeks and were below baseline values at six months, but rates of bone resorption, and blood levels of NTX, remained elevated for six months in this model.[89]

More recently, a mouse model of SCI-induced bone loss has been characterized by Picard and colleagues.[90] In male mice studied one month after spinal cord transection at the low-thoracic level (T9–10), a decrease in femoral bone volume (−22%), trabecular thickness (−10%), and trabecular number (-14%) was observed. DXA measurements revealed no change in BMD but a significant reduction (−14%) in bone mineral content. With an appreciation of the increasing availability of genetic and molecular research tools for investigation in mice, the murine model may be useful to further define the cellular and molecular mechanisms of demineralization associated with SCI.[90]

## Endocrine influences on muscle and bone after SCI

### Muscle

**Endogenous and exogenous glucocorticoids and other catabolic hormones.** Catabolic influences present during the first few days or weeks after SCI may exacerbate the effects of immobilization on muscle atrophy. These include reduced caloric intake, infection, and the administration of glucocorticoid medications. The latter is of particular concern because, in an attempt to improve functional outcome, "megadoses" of methylprednisolone that are approximately 30-fold greater than those used for any other clinical indication are administered for 24–48 h, beginning at the time of SCI.[91–97] A decrease in muscle mass and strength, as well as the development of glucocorticoid myopathies, are well-recognized consequences of glucocorticoid administration, and it has been suggested that even a brief exposure to the high doses of methylprednisolone administered to improve neurological outcomes after SCI could severely impair muscle function.[98] In paraplegic rats administered this regimen of methylprednisolone for 24 h, triceps muscle weights 7 days after SCI (or 6 days after cessation of glucocorticoid administration) were reduced by approximately 50% compared to those administered vehicle (Cardozo *et al.*, in press). In a case series examining effects of high-dose methylprednisolone on neuromuscular function, five of six individuals administered this agent demonstrated histopathological evidence of glucocorticoid myopathy consisting of necrotic muscle fibers associated with electromyelographic evidence of neuromuscular dysfunction that persisted for months.[98] Elevated levels of endogenous glucocorticoids have been reported to contribute to muscle atrophy in some pathophysiological states, but serum cortisol levels, and their diurnal variation, in healthy persons with chronic SCI, have been reported to be normal.[99,100]

Elevation in the levels of angiotensin II may also contribute to muscle atrophy in individuals with higher cord lesions. Angiotensin II stimulates muscle catabolism in animals and has been linked to cardiac cachexia.[101,102] In persons with tetraplegia in whom sympathetic regulation of vascular tone, heart rate, and cardiac contractility are absent or impaired, alternative mechanisms must be recruited to prevent orthostatic hypotension. Foremost amongst these is activation of the renin–angiotensin–aldosterone axis, with increased angiotensin II levels, and associated vasoconstriction and salt and water retention.[103] The degree to which this adaptive vascular homeostatic mechanism contributes to muscle wasting after SCI remains unknown.

**Testosterone.** Reductions in local or circulating levels of anabolic hormones may accentuate skeletal muscle atrophy in persons in the general population, and this predisposition may be assumed to be operative as well in persons with immobilization. It is well appreciated that muscle mass and strength are increased by testosterone.[104,105] Suppression of pituitary release of gonadotropins decreases testosterone levels, with potential negative effects on skeletal muscle mass and function.[106] In cross-sectional studies, individuals with SCI had reduced serum testosterone levels when compared to controls,[11,107] and over one-third had clinically significant reductions in testosterone.[11] These changes may be of greatest relevance to preservation of mass and function of skeletal muscle in individuals with some preservation of neurological function below the level of lesion. Testosterone, or its synthetic derivatives, nandrolone and oxandrolone, increase muscle mass in individuals with HIV infection,[108–110] COPD,[111,112] or burn injury,[113,114] and increase muscle mass and strength in individuals administered glucocorticoids.[115] In animal mod-

els, testosterone, or its synthetic derivative, nandrolone, reduce muscle atrophy due to SCI, nerve transection, immobilization, unweighting, or glucocorticoid administration.[25,116–119] A clinical trial is currently being performed to determine whether testosterone replacement therapy improves muscle mass and/or function in hypogonadal men with chronic SCI (WA Bauman, unpublished).

Studies in humans and animal models have provided insight into the mechanisms by which testosterone reduces and/or reverses muscle atrophy. In healthy individuals, testosterone exhibits anabolic activity, promoting the retention of nitrogen and increasing skeletal muscle mass. Conversely, reductions in testosterone levels diminish muscle mass associated with reduced uptake and delayed clearance of leucine, reflecting reduced muscle protein synthetic rates.[106] Similarly, in healthy young men, oxandrolone, an anabolic steroid, increased muscle protein synthesis without altering muscle protein breakdown.[120] Protein synthetic rates are controlled by the protein kinase mTOR (mammalian target of rapamycin), through phosphorylation of its downstream substrates eIF4E binding protein (4EBP) and p70S6 kinase.[121] Activity of mTOR has been found to be necessary for muscle hypertrophy due to overloading, and for recovery of muscle mass after immobilization.[37] In the highly androgen-responsive levator ani muscle, administration of androgens to castrated rats resulted in increased phosphorylation of p70S6 kinase at an mTOR-dependent site, suggesting that testosterone and other androgens activate mTOR.[122]

Androgens may exert this action via upregulation of IGF-1 and subsequent activation of mTOR through IGF-1 receptor/PI3K/Akt signaling. IGF-1 increases the size of skeletal muscle through mTOR-dependent pathways[123] and prevents muscle loss in several models of muscle atrophy of differing etiologies.[124–126] The proximal IGF-1 promotor has been found to contain two androgen response elements.[127] A fall in testosterone levels reduced muscle IGF-1 mRNA levels,[106] whereas testosterone administration to hypogonadal, elderly men increased IGF-1 protein,[128] and testosterone replacement therapy increased Akt phosphorylation in patients infected with HIV.[110] Nevertheless, a causal relationship between changes in IGF-1 expression and testosterone actions on skeletal muscle remains to be established.

Activity of mTOR is inhibited by "regulated in development and DNA damage responses" (REDD) – 1 and its homolog REDD2, which originally were identified as stress, hypoxia, and DNA damage-inducible genes.[129] The findings that REDD1 is markedly upregulated in skeletal muscle by dexamethasone linked this protein to the well-known actions of glucocorticoids to inhibit mTOR and muscle protein synthesis.[130] Testosterone has been shown to block the adverse effects of glucocorticoids on REDD1 expression and mTOR activity.[131] Testosterone reduces muscle protein catabolism in burn victims[132] and in cultured myoblasts treated with dexamethasone.[118] Testosterone also reduces expression of MAFbx in myoblasts, and attenuates upregulation of MAFbx and MuRF1 by glucocorticoids.[118] Nandrolone reduces expression of these ubiquitin ligases resulting from paralysis due to nerve transection.[25]

**IGF-1.** In animal models of muscle atrophy from unweighting, starvation, diabetes, or glucocorticoid administration, muscle loss can be prevented by IGF-1.[124–126] Through activation of PI3K/Akt, IGF-1 regulates multiple intracellular signals that inhibit muscle atrophy programs. Foremost among these is the phosphorylation and inactivation of the transcription factors FOXO1 and FOXO3A, and consequently, reduction in expression of MAFbx and MuRF1.[133,134] The growth hormone (GH)/IGF-1 axis appears to be depressed in many individuals with SCI. Basal levels of plasma IGF-1 were lower in younger persons with chronic SCI than age-matched able-bodied controls, while no difference could be appreciated in older individuals;[11,107] arginine-induced release of GH was blunted in persons with SCI, regardless of age, suggesting that function of the GH/IGF-1 axis was depressed. Administration of low-doses baclofen, a gamma-aminobutyric acid (GABA) B antispasmotic medication, was found to increase plasma IGF-1 levels.[135] Paradoxically, in an animal model of SCI, muscle mRNA levels for IGF-1 and its receptor were elevated during the first week after SCI, although the biological significance of these findings is uncertain.[19] One possibility is that such changes reflect an unsuccessful countermeasure to the catabolic changes elicited by paralysis.

**Muscle loss as a factor in insulin resistance after SCI.** Cross sectional studies have shown that about half of persons with paraplegia, and two-thirds of individuals with tetraplegia, have an abnormal oral glucose tolerance test or diabetes.[7,8] Skeletal muscle is responsible for up to 75% of insulin-stimulated clearance of glucose from blood, which is greatly increased by physical activity [for recent reviews, see Refs. 136 and 137]. Consequently, the inactivity and marked muscle atrophy in SCI has been assumed to contribute to the increased propensity for disorders of carbohydrate metabolism. Consistent with this view, exercise training of paralyzed muscles of the lower extremities using FES three times a week for 1 year significantly improved insulin sensitivity.[138] Additionally, disturbances of glucose and insulin sensitivity are more common in persons with tetraplegia,[7,8] in whom the greatest loss of skeletal muscle are observed[139,140] and levels of activity are lowest.

Individuals with SCI have increased intramuscular fat by MRI, which on histopathology reflects both intracellular fat and fat between muscle fibers.[141,142] In sedentary able-bodied individuals, increased intramuscular fat correlates negatively with insulin sensitivity,[143] and a similar relationship is observed in individuals with SCI.[141] The causal role of increased intramuscular fat in insulin insensitivity may be questioned by the finding that lean, highly conditioned individuals also have increased intramuscular fat.[144] Thus, it may be that after SCI, increased intramuscular fat is a reflection of more fundamental aberrations in metabolism of paralyzed muscle.

The possibility that myokines liberated during muscle atrophy may affect distant tissues, in a manner similar to adipokines, is suggested by findings that in some states of neurological impairment predisposing to muscle atrophy, such as stroke,[145] or in mice with muscle-restricted deletion of the PGC-1α gene,[18] muscle releases the cytokine TNF. Little is known as to whether such myokines alter the biology of nearby or distant bone or fat, or have other influences on the function of other tissues and organs, but this remains an interesting question for further investigation.

### Bone

**Estrogen.** It is well appreciated that BMD, even in men, is more closely related to serum estradiol concentrations, especially the bioactive form, than to serum testosterone concentrations[146,147] [for review, see Refs. 148-151]. Serum estrogen concentrations in men are largely dependent upon conversion of androgen precursors to estradiol and estrone by peripheral aromatase activity.[152,153] It also appears that estrogen hydroxylation is related to BMD in men; 16α hydroxyestrone (16α OHE) and estriol were positively correlated with adjusted BMD in all regions of the proximal femur, and men in the highest tertile of urinary 16-α OHE had the highest BMD.[154] The Osteoporotic Fractures in Men Study was recently performed in a large population of community-dwelling older men. Men with low bioactive estradiol (<11.4 pg/mL) or high sex-hormone binding globulin (>59.1 nm) levels were at higher risk for a non-vertebral fracture; in addition, men with low bioactive testosterone and high SHBG had a higher fracture risk, with the predictive value strongest when all measures were considered in combination.[155] Studies that examine changes in levels of serum estrogen in men after acute SCI and their correlation with subsequent changes in BMD and risk of fracture have not been performed. Due to the reported depression in serum testosterone after acute SCI, levels of estrogenic compounds would be expected to be diminished.

The administration of high levels of methylprednisolone in an attempt to preserve neurological function immediately after SCI[97] is still fairly widely practiced in clinical medicine. Estrogen has been shown to have a protective effect against osteocyte apoptosis by antagonizing the translocation of the glucocorticoid receptor to the nucleus.[156] In a large population study by gene-wide SNP association analysis, regulation of BMD was associated with a single-nucleotide polymorphism (SNP) of the SOST gene in its upstream enhancer region, which also contained putative estrogen receptor α binding sites.[157]

**Testosterone.** Testosterone is anabolic to bone. The androgen receptor (AR) is found in all three bone cells: osteoblasts, osteoclast, and osteocytes.[158,159] Androgens may promote proliferation and differentiation of osteoblasts,[160] inhibit osteoclast recruitment and bone resorption activity,[161] or affect osteoblast-to-osteoclast signaling, suggesting a role for AR in regulating bone remodeling.[73,160] In men with idiopathic osteoporosis, the decrease in testosterone levels correlated with a decline in BMD, and testosterone replacement therapy increased BMD in the vertebrae [for review see Ref. 159]. In rat mod-

els of bone loss due to hind limb unweighting, administration of either testosterone or nandrolone blocked much of the immobilization-related decrease in BMD.[116] Recently, we tested whether nandrolone protected bone against loss in male rats that underwent sciatic nerve transection, followed 28 days later by treatment with nandrolone or vehicle for 28 days. Denervation led to reductions in BMD of 7 and 12% for femur and tibia, respectively. Nandrolone preserved 80% and 60% of BMD in femur and tibia, respectively, demonstrating that nandrolone administration significantly reduced loss of BMD due to denervation.[162] The possibility of a similar beneficial role of androgens in prevention of SCI-induced bone loss in animal models is under investigation in our laboratory. Clinical trials are required to determine whether testosterone replacement therapy preserves bone mass in hypogonadal men with SCI.

**Calcium homeostasis, parathyroid hormone, and vitamin D.** Because of increased bone resorption after acute immobilization, calciuria increases within days after injury, becoming maximal between one and six months.[163–166] Hypercalcemia usually does not occur in adults after acute SCI who have normal renal function, but may be observed in those with impaired fractional excretion of calcium by the kidney, which may occur in childhood, as a consequence of acute renal insufficiency at time of acute initial presentation, or underlying chronic renal disease. Persons who have higher bone turnovers, such as patients who present with multiple bone fractures, Paget's disease, or children with acute SCI, may be particularly susceptible to hypercalcemia.[167] Other risk factors that have been found to predispose to hypercalcemia include recent paralysis, male gender, complete neurologic injuries, high cervical injury, dehydration, and prolonged immobilization.[165] Maximum urinary calcium excretion in those with SCI is between two and four times that of able-bodied subjects who were placed at prolonged, voluntarily bed rest. In most reports,[164] but not all,[168] urinary calcium excretion was not reduced by weight-bearing exercise or wheelchair activity.

The parathyroid hormone (PTH)–vitamin D axis is suppressed following acute SCI due to increased levels of circulating ionized calcium, with appropriately suppressed PTH release associated with depressed levels of 1,25-dihydroxyvitamin D

[1,25(OH)$_2$D] and nephrogenous cyclic AMP.[169] Stewart and colleagues[169] found that patients within the initial two to three months after acute SCI, who were either on a restricted (400 mg calcium) or unrestricted (1,160 mg calcium) calcium intake, had similar degrees of hypercalciuria and serum calcium levels, suggesting markedly reduced gut calcium absorption due to the observed suppression of the PTH-vitamin D axis. In a longitudinal study of patients after acute SCI, a decrease in serum PTH levels was initially noted at three weeks.[80] Mechanick and colleagues[170] found that patients with complete SCI had a greater suppression of the PTH—vitamin D axis than those with incomplete injury, suggesting increased degrees of bone resorption in patients with increasing neurological impairment.

The finding of hypercalciuria and occasionally hypercalcemia after acute SCI has led many health care providers to the clinical practice of dietary restriction of calcium, and by association, vitamin D. The practice of restricting dietary intake of calcium at time of acute SCI, and frequently thereafter, is not directed at the pathogenesis of this disorder, which is increased bone resorption due to immobilization.[169] The dietary vitamin D restriction with resulting deficiency state may be worsened by reduced sunlight exposure[171] and prescription medications, including anticonvulsants, that induce hepatic microsomal enzymes, which accelerate vitamin D metabolism.[172,173] The combination of reduced calcium intake and a relative vitamin D deficiency state may lower the ionized serum calcium concentration, resulting in a state of secondary hyperparathyroidism.

In 100 subjects with chronic SCI (mean duration of injury of 20 years) compared with 50 normal control subjects, those with SCI had significantly lower 25-hydroxyvitamin D [25(OH)D] levels, the major storage form of vitamin D; serum 25(OH)D negatively correlated with serum PTH.[174] Serum 1,25(OH)$_2$D levels were significantly higher in persons with chronic SCI, with several patients having levels above the normal range; serum 1,25(OH)$_2$D levels were positively correlated with plasma PTH levels.[174]

PTH has long been appreciated to be a proresorptive hormone, as illustrated by findings that PTH-null mice have a high bone mass.[175] Thus, a state of secondary hyperparathyroidism in persons with chronic SCI would be expected to be associated with increased bone turnover and eventual bone loss.[176,177] As a proof of this concept, in persons with chronic SCI and relative vitamin D deficiency [25(OH)D ≤ 20 mg/ml] and/or a high PTH (>55 pg/mL), a calcium infusion decreased levels of NTX, a marker of osteoclastic activity.[178] However, it is now well appreciated, and currently employed in clinical practice, that the intermittent administration of PTH is anabolic to bone by shifting osteoblasts away from RANKL production toward osteogenesis.[73] Although of possible clinical value in persons with acute or chronic SCI, studies administering PTH$_{1–34}$ (e.g., teriparatide) as a potential bone-anabolic agent have not been reported to date.

## Neurological dysfunction after SCI contributes to loss of muscle and bone

### Muscle

After SCI, the spinal cord is electrically silent, a state referred to as spinal shock. Within several days, electrical activity has resumed, and by two weeks, reflex circuits have again become active, resulting in involuntary electrical activity of motor neurons which varies in intensity depending on the level and completeness of the injury.[12] As might be expected, the presence and degree of spasticity appear to correlate with muscle mass. In a case series in which severe spasticity was treated with a myelotomy, muscle biopsy samples revealed decreased muscle fiber cross-sectional area at two months after myelotomy.[179] In a cross sectional study of individuals with incomplete SCI, spasticity positively correlated with muscle mass at six weeks after injury.[72] In an animal model of SCI at 7 days after injury, in which reflex arcs are disrupted surgically, electrical stimulation of limb muscles for brief periods (4 sec) several times a day was sufficient to prevent atrophy, as well as prevent the upregulation of MAFbx and MuRF1.[28]

Emerging evidence indicates that patterns of residual neurological function after SCI are far more complex than the elicitation of simple reflexes. The spinal cord contains neural networks capable of initiating and controlling coordinated stepping that can be stimulated by sensory inputs, epidural electrical stimulation or pharmaceuticals.[180] The degree to which these circuits can be recruited to support body weight and appropriately load skeletal muscle to mitigate atrophy and/or fiber type change remains uncertain, but is now a topic of study at several centers.

A fundamental question has been the degree to which the lower motor neuron itself influences the biology of skeletal muscle during immobilization. Studies examining this question have compared changes elicited by paralysis due to nerve transection with those resulting from spinal isolation. In female rats at 14 days after spinal isolation, less atrophy of hindlimb muscles was observed compared to sciatic nerve transection. The spinal isolation animals also demonstrated rapid reductions in initial elevations in expression of the myogenic factors, MyoD and myogenin.[181,182] It has thus been argued that innervation of skeletal muscle by lower motor neurons directly influences muscle biology, even in the absence of neuromuscular activity. These findings have similarities with those from a study of the comparison of gene expression changes resulting from nerve transection or transection of the thoracic spinal cord in male rats, in which muscle from spinal cord injured animals displayed more complete return to normal levels of the expression of MAFbx and MuRF1.[19] In this study, atrophy of hindlimb muscles was similar for both denervated and spinal cord injured rats at 14 days. Of note, in a comparison of data from the studies of the female rats with spinal isolation and the male rats with SCI, there was a greater percentage of muscle lost in the latter model.[181,182] The reasons for this finding are not clear, but gender differences in response to neurological injury and/or sex-steroids offer possible explanations.

## Bone

Both sympathetic and sensory nerve fibers are present in the periosteum, bone marrow and mineralized bone.[76] Several lines of evidence suggest that loss of sympathetic nervous system (SNS) outflow from the SCI at vertebral levels above those providing preganglionic neurons to the sympathetic chain may influence bone loss due to immobilization. The SNS has been found to modulate bone through ß-adrenergic receptors[183] with leptin and neuromedin U playing key roles in central control of SNS activity.[183,184] Leptin has been reported to centrally influence sympathetic output, with lower levels causing reduction of ß2 adrenergic receptor activation on osteoblasts and high bone mass.[183] Various neuropeptides are found in bone, including calcitonin gene-related peptide, substance

P, vasoactive intestinal peptide, and neuropeptide Y.[185,186] Of note, these neuropeptides may modulate the OGP/RANKL/RANK system.[187,188] Furthermore, there is evidence that bone remodeling may be under circadian sympathetic regulation via clock genes that inhibit osteoblast proliferation [for review, see[189]], which may be disrupted after lesions of the high thoracic and cervical regions. Definitive studies that clearly define the role of the sensory and sympathetic innervation of bone after SCI, or other forms of neurological insult, have not yet been performed.

## Vascular adaptations

Femoral artery blood flow and diameter are reduced in persons with SCI, whereas blood flow per unit volume of preserved muscle is normal.[190] The possibility that these alterations limit exercise performance and explain, in part, the easy fatigability of individuals with chronic SCI has been explored through studies of blood flow and FES-induced resistance training.[26] Twice-weekly resistance training significantly increased muscle mass and reduced fatigability significantly, but did not appear to alter measures of peak femoral artery flow or diameter after exertion, or the period of hyperemia after ischemia. While blood flow *per se* appears appropriate to tissue volumes, vascular reactivity is nonetheless impaired after SCI. In a study of individuals with incomplete SCI (paresis rather than paralysis), the period of hyperemia after ischemia produced with a cuff around the calf was prolonged compared to able-bodied controls.[190] Whether this results from inactivity, disruption of sympathetic innervation of the vasculature, elevated circulating levels of angiotensin, or other abnormalities is not known. Regardless of the mechanism, the altered vascular reactivity observed after SCI suggests the possibility that distribution of blood through the muscular capillary bed may be abnormal in individuals with SCI due to impaired local vasoregulatory influences in response to exertion, analogous to shunting in the pulmonary or hepatic vascular bed.

## Interactions of fat, muscle, and bone after SCI

Body composition markedly deteriorates during the first six months after SCI, with profound loss of bone and muscle, as described above, and marked gains in adipose tissue.[27,191,192] A recent study reported that patients aged 10–21 years with SCI have significantly decreased lean tissue mass, decreased bone mineral content, and increased fat mass in those with paraplegia.[193] Interestingly, in a study of identical twins discordant for SCI, a strong relationship was observed between total and regional body fat and BMD of the legs of the twins with SCI, but not in their able-bodied cotwins. A potential causal relationship between increased fat and bone mass was suggested by findings in twins with SCI of a positive correlation between serum estradiol concentration and leg BMD.[61]

Osteoblasts and adipocytes are derived from a common bone marrow progenitor cell, a mesenchymal stem cell. Intriguingly, increases in bone marrow fat are reported after immobilization, including SCI. In a study in which young healthy women were put to bed rest for 60 days, bone marrow fat of the lumbar vertebrae increased on average by 2.5%.[194] In patients with paraplegia, transiliac bone biopsies revealed an increase of bone marrow adipose volume of about 36% at 12 weeks after SCI.[195] Bone marrow fat was also increased in rats after spaceflight.[196] Accumulation of bone marrow fat may be detrimental to bone health. In a rodent model of weightlessness by hindlimb suspension, accumulation of bone marrow fat was found to be associated with a diminished rate of longitudinal bone growth, reduced mass of mineralized tissue, and a reduced number of osteoblasts.[197]

Molecular events that determine the shift in lineage of mesenchymal stem cells toward formation of bone marrow fat in states of disuse remain uncertain. One possibility is loss of mechanical loading of these cells, as suggested by findings that in cell culture, cellular strain favored differentiation of progenitors into an osteoblast-like cell.[78] Another potential factor is loss of the influence of sex steroids (e.g., estrogenic and androgenic) on progenitor cells, as suggested by the significant prevalence of relative or absolute hypogonadism in those with SCI,[11,107,198] and by findings that androgens inhibit adipogenesis of 3T3-L1 cells in a cell culture system.[199] Growth hormone deficiency in persons with SCI may also play a role in determining bone marrow stem cell fate, with a deficiency state predisposing to adipogenesis. In hypophysectomized rodents, increased bone marrow adiposity was observed, although paradoxically, adipocyte precursor cell pools were diminished; these alterations were normalized with GH replacement.[200]

Adipogenic differentiation factors may also play critical roles in alterations in bone marrow fat during immobilization, including those occurring as a result of SCI. Activity of the peroxisome proliferator activated receptor (PPAR)-γ is necessary and sufficient for adipogenic differentiation.[16,201] The possibility that administration of high doses of methylprednisone at time of acute SCI may also tend to drive precursor stem cells to adipogenesis is suggested by findings that glucocorticoids promote upregulation of PPARγ.[202] That mechanical loading of precursor cells regulates signals that oppose the effects of PPARγ is illustrated in a tissue culture model of primary stromal cells. Cyclic loading induced higher Runx2 and lower PPARγ; cyclic stretching was able to compensate for treatment with a PPARγ activator and partially overcome the induction of adipogenesis.[203] Adverse effects of glucocorticoids on progenitor cell fate may be exacerbated by actions of this class of steroids to inhibit osteoblast cell cycle progression and function [for review, see Refs. 202 and 204].

Adipose tissue releases numerous soluble mediators (adipokines) that include leptin and adiponectin as well as inflammatory cytokines.[205] These adipocyte-derived molecules have been suggested to provide autocrine and paracrine regulation of adipocytes, promote a state of systemic inflammation, and regulate fat metabolism, insulin sensitivity, and indirectly, through leptin-dependent modulation of the sympathetic nervous system, bone biology.[183,184,205–207] Changes in adipokine levels after SCI have not been completely evaluated, but appear to be related to total body fat mass. In an early cross-sectional study, changes in levels of leptin after SCI correlated well with fat mass after SCI.[208] In a study of 44 individuals with SCI, leptin and plasminogen activator inhibitor-1 levels were related to measures of abdominal fat, while levels of adiponectin were inversely correlated.[209] It remains possible that adipokines produced by bone marrow fat influence the fate or biology of nearby bone marrow mesenchymal stem cells and/or bone cells, although this possibility has not been studied.

## Acknowledgments

The research reported here was supported by the Veterans Health Administration, Rehabilitation

Research and Development Service (B4162C, B3347K).

## Conflicts of interest

The authors declare no conflicts of interest.

## References

1. Acton, P.A. *et al.* 1993. Traumatic spinal cord injury in Arkansas, 1980 to 1989. *Arch. Phys. Med. Rehabil.* **74:** 1035–1040.

2. Guertin, P.A. 2008. A technological platform to optimize combinatorial treatment design and discovery for chronic spinal cord injury. *J. Neurosci. Res.* **86:** 3039–3051.

3. Berkowitz, M. (ed.) 1998. *Spinal Cord Injury—An Analysis of Medical and Social Costs.* Vol. 107, Demos Medical Publishing. New York.

4. Dudley-Javoroski, S. & R.K. Shields. 2008. Muscle and bone plasticity after spinal cord injury: review of adaptations to disuse and to electrical muscle stimulation. *J. Rehabil. Res. Dev.* **45:** 283–296.

5. Biering-Sorensen, B., I.B. Kristensen, M. Kjaer & F. Biering-Sorensen. 2009. Muscle after spinal cord injury. *Muscle Nerve* **40:** 499–519.

6. Jiang, S.D., L.Y. Dai & L.S. Jiang. 2006. Osteoporosis after spinal cord injury. *Osteoporos Int.* **17:** 180–192.

7. Bauman, W.A., R.H. Adkins, A.M. Spungen & R.L. Waters. 1999. The effect of residual neurological deficit on oral glucose tolerance in persons with chronic spinal cord injury. *Spinal Cord* **37:** 765–771.

8. Bauman, W.A. & A.M. Spungen. 1994. Disorders of carbohydrate and lipid metabolism in veterans with paraplegia or quadriplegia: a model of premature aging. *Metabolism* **43:** 749–756.

9. Bauman, W.A. *et al.* 1992. Coronary artery disease: metabolic risk factors and latent disease in individuals with paraplegia. *Mt. Sinai J. Med.* **59:** 163–168.

10. Bauman, W.A. *et al.* 1992. Depressed serum high density lipoprotein cholesterol levels in veterans with spinal cord injury. *Paraplegia* **30:** 697–703.

11. Tsitouras, P.D., Y.G. Zhong, A.M. Spungen & W.A. Bauman. 1995. Serum testosterone and growth hormone/insulin-like growth factor-I in adults with spinal cord injury. *Horm. Metab. Res.* **27:** 287–292.

12. Adams, M.M. & A.L. Hicks. 2005. Spasticity after spinal cord injury. *Spinal Cord* **43:** 577–586.

13. Mathias, C.J. & H.L. Frankel. 1988. Cardiovascular control in spinal man. *Annu. Rev. Physiol.* **50:** 77–592.

14. Bickel, C.S., J.M. Slade & G.A. Dudley. 2004. Long-term spinal cord injury increases susceptibility to isometric contraction-induced muscle injury. *Eur. J. Appl. Physiol.* **91:** 308–313.

15. Lin, J. *et al.* 2002. Transcriptional co-activator PGC-1 alpha drives the formation of slow-twitch muscle fibres. *Nature* **418:** 797–801.

16. Rosen, E.D. & B.M. Spiegelman. 2001. PPARgamma : a nuclear regulator of metabolism, differentiation, and cell growth. *J. Biol. Chem.* **276:** 37731–37734.

17. Sacheck, J.M. *et al.* 2007. Rapid disuse and denervation atrophy involve transcriptional changes similar to those of muscle wasting during systemic diseases. *FASEB J.* **21:** 140–155.

18. Handschin, C. *et al.* 2007. Skeletal muscle fiber-type switching, exercise intolerance, and myopathy in PGC-1alpha muscle-specific knock-out animals. *J. Biol. Chem.* **282:** 30014–30021.

19. Zeman, R.J. *et al.* 2009. Differential skeletal muscle gene expression after upper or lower motor neuron transection. *Pflugers Arch.* **458:** 525–535.

20. Urso, M.L. *et al.* 2007. Alterations in mRNA expression and protein products following spinal cord injury in humans. *J. Physiol.* **579:** 877–892.

21. Clarke, B.A. *et al.* 2007. The E3 Ligase MuRF1 degrades myosin heavy chain protein in dexamethasone-treated skeletal muscle. *Cell Metab.* **6:** 376–385.

22. Bodine, S.C. *et al.* 2001. Identification of ubiquitin ligases required for skeletal muscle atrophy. *Science* **294:** 1704–1708.

23. Sandri, M. *et al.* 2006. PGC-1alpha protects skeletal muscle from atrophy by suppressing FoxO3 action and atrophy-specific gene transcription. *Proc. Natl. Acad. Sci. USA* **103:** 16260–16265.

24. Wang, X. *et al.* 2005. Runx1 prevents wasting, myofibrillar disorganization, and autophagy of skeletal muscle. *Genes Dev.* **19:** 1715–1722.

25. Zhao, J. *et al.* 2008. Effects of nandrolone on denervation atrophy depend upon time after nerve transection. *Muscle Nerve* **37:** 42–49.

26. Sabatier, M.J. *et al.* 2006. Electrically stimulated resistance training in SCI individuals increases muscle fatigue resistance but not femoral artery size or blood flow. *Spinal Cord* **44:** 227–233.

27. Baldi, J.C., R.D. Jackson, R. Moraille & W.J. Mysiw. 1998. Muscle atrophy is prevented in patients with acute spinal cord injury using functional electrical stimulation. *Spinal Cord* **36:** 463–469.

28. Kim, S.J. *et al.* 2008. Gene expression during inactivity-induced muscle atrophy: effects of brief bouts of a forceful contraction countermeasure. *J. Appl. Physiol.* **105:** 1246–1254.

29. Adams, G.R. 2002. Autocrine and/or paracrine insulin-like growth factor-I activity in skeletal muscle. *Clin. Orthop. Relat. Res.*, **403** Suppl. S188–S196.

30. Lange, S. *et al.* 2005. The kinase domain of titin controls muscle gene expression and protein turnover. *Science* **308:** 1599–1603.

31. Sakamoto, K. & L.J. Goodyear. 2002. Invited review: intracellular signaling in contracting skeletal muscle. *J. Appl. Physiol.* **93:** 369–383.

32. Semsarian, C. *et al.* 1999. Skeletal muscle hypertrophy is mediated by a Ca2+-dependent calcineurin signalling pathway. *Nature* **400:** 576–581.

33. Schiaffino, S. & A. Serrano. 2002. Calcineurin signaling and neural control of skeletal muscle fiber type and size. *Trends Pharmacol. Sci.* **23:** 569–575.

34. Bassel-Duby, R. & E.N. Olson. 2006. Signaling pathways in skeletal muscle remodeling. *Annu. Rev. Biochem.* **75:** 19–37.

35. Dunn, S.E., J.L. Burns & R.N. Michel. 1999. Calcineurin is required for skeletal muscle hypertrophy. *J. Biol. Chem.* **274:** 21908–21912.

36. Dunn, S.E., E.R. Chin & R.N. Michel. 2000. Matching of calcineurin activity to upstream effectors is critical for skeletal muscle fiber growth. *J. Cell Biol.* **151:** 663–672.

37. Bodine, S.C. *et al.* 2001. Akt/mTOR pathway is a crucial regulator of skeletal muscle hypertrophy and can prevent muscle atrophy in vivo. *Nat. Cell Biol.* **3:** 1014–1019.

38. Dupont-Versteegden, E.E., M. Knox, C.M. Gurley, *et al.* 2002. Maintenance of muscle mass is not dependent on the calcineurin-NFAT pathway. *Am. J. Physiol. Cell Physiol.* **282:** C1387–C1395.

39. Li, H.H. *et al.* 2004. Atrogin-1/muscle atrophy F-box inhibits calcineurin-dependent cardiac hypertrophy by participating in an SCF ubiquitin ligase complex. *J. Clin. Invest.* **114:** 1058–1071.

40. Akimoto, T. *et al.* 2005. Exercise stimulates Pgc-1alpha transcription in skeletal muscle through activation of the p38 MAPK pathway. *J. Biol. Chem.* **280:** 19587–19593.

41. Calabria, E. *et al.* 2009. NFAT isoforms control activity-dependent muscle fiber type specification. *Proc. Natl. Acad. Sci. USA* **106:** 13335–13340.

42. Mohr, T. *et al.* 1997. Long-term adaptation to electrically induced cycle training in severe spinal cord injured individuals. *Spinal Cord* **35:** 1–16.

43. Giangregorio, L. & N. McCartney. 2006. Bone loss and muscle atrophy in spinal cord injury: epidemiology, fracture prediction, and rehabilitation strategies. *J. Spinal Cord Med.* **29:** 489–500.

44. Szollar, S.M., E.M. Martin, D.J. Sartoris, *et al.* 1998. Bone mineral density and indexes of bone metabolism in spinal cord injury. *Am. J. Phys. Med. Rehabil.* **77:** 28–35.

45. Garland, D.E., R.H. Adkins, V. Kushwaha & C. Stewart. 2004. Risk factors for osteoporosis at the knee in the spinal cord injury population. *J. Spinal Cord Med.* **27:** 202–206.

46. Warden, S.J. *et al.* 2002. Quantitative ultrasound assessment of acute bone loss following spinal cord injury: a longitudinal pilot study. *Osteoporos. Int.* **13:** 586–592.

47. Vico, L. *et al.* 2000. Effects of long-term microgravity exposure on cancellous and cortical weight-bearing bones of cosmonauts. *Lancet* **355:** 1607–1611.

48. Leblanc, A.D., V.S. Schneider, H.J. Evans, *et al.* 1990. Bone mineral loss and recovery after 17 weeks of bed rest. *J. Bone Miner. Res.* **5:** 843–850.

49. Berarducci, A. 2009. Stopping the silent progression of osteoporosis. *Am. Nurse Today* **3:** 18.

50. Eser, P. *et al.* 2004. Relationship between the duration of paralysis and bone structure: a pQCT study of spinal cord injured individuals. *Bone* **34:** 869–880.

51. Bauman, W.A., A.M. Spungen, J. Wang, *et al.* 1999. Continuous loss of bone during chronic immobilization: a monozygotic twin study. *Osteoporos. Int.* **10:** 123–127.

52. de Bruin, E.D., V. Dietz, M.A. Dambacher & E. Stussi. 2000. Longitudinal changes in bone in men with spinal cord injury. *Clin. Rehabil.* **14:** 145–152.

53. Dauty, M., B. Perrouin Verbe, Y. Maugars, *et al.* 2000. Supralesional and sublesional bone mineral density in spinal cord-injured patients. *Bone* **27:** 305–309.

54. Biering-Sorensen, F., H.H. Bohr & O.P. Schaadt. 1990. Longitudinal study of bone mineral content in the lumbar spine, the forearm and the lower extremities after spinal cord injury. *Eur. J. Clin. Invest.* **20:** 330–335.

55. Finsen, V., B. Indredavik & K.J. Fougner. 1992. Bone mineral and hormone status in paraplegics. *Paraplegia* **30:** 343–347.

56. Frey-Rindova, P., E.D. de Bruin, E. Stussi, *et al.* 2000. Bone mineral density in upper and lower extremities during 12 months after spinal cord injury measured by peripheral quantitative computed tomography. *Spinal Cord* **38:** 26–32.

57. Eser, P., A. Frotzler, Y. Zehnder & J. Denoth. 2005. Fracture threshold in the femur and tibia of people with spinal cord injury as determined by peripheral quantitative computed tomography. *Arch. Phys. Med. Rehabil.* **86:** 498–504.

58. Zehnder, Y. *et al.* 2004. Long-term changes in bone metabolism, bone mineral density, quantitative ultrasound parameters, and fracture incidence after spinal cord injury: a cross-sectional observational study in 100 paraplegic men. *Osteoporos. Int.* **15:** 180–189.

59. Morse, L.R. *et al.* 2009. Osteoporotic fractures and hospitalization risk in chronic spinal cord injury. *Osteoporos. Int.* **20:** 385–392.

60. Zeman, R.J., A. Hirschman M.L. Hirschman. 1991. Clenbuterol, a beta 2-receptor agonist, reduces net bone loss in denervated hindlimbs. *Am. J. Physiol.* **261:** E285–E289.

61. Bauman, W.A., A.M. Spungen, J. Wang, *et al.* 2006. Relationship of fat mass and serum estradiol with lower extremity bone in persons with chronic spinal cord injury. *Am. J. Physiol. Endocrinol. Metab.* **290:** E1098–E1103.

62. Shields, R.K. & S. Dudley-Javoroski. 2006. Musculoskeletal plasticity after acute spinal cord injury: effects of long-term neuromuscular electrical stimulation training. *J. Neurophysiol.* **95:** 2380–2390.

63. Shields, R.K., S. Dudley-Javoroski & L.A. Law. 2006. Electrically induced muscle contractions influence bone density decline after spinal cord injury. *Spine (Phila Pa 1976)* **31:** 548–553.

64. Dudley-Javoroski, S. & R.K. Shields. 2008. Asymmetric bone adaptations to soleus mechanical loading after spinal cord injury. *J. Musculoskelet. Neuronal Interact* **8:** 227–238.

65. Shields, R.K. & S. Dudley-Javoroski. 2007. Musculoskeletal adaptations in chronic spinal cord injury: effects of long-term soleus electrical stimulation training. *Neurorehabil. Neural Repair* **21:** 169–179.

66. Mohr, T. *et al.* 1997. Increased bone mineral density after prolonged electrically induced cycle training of paralyzed limbs in spinal cord injured man. *Calcif. Tissue Int.* **61:** 22–25.

67. Goktepe, A.S., I. Tugcu, B. Yilmaz, *et al.* 2008. Does standing protect bone density in patients with chronic spinal cord injury? *J. Spinal Cord Med.* **31:** 197–201.

68. Giangregorio, L.M. *et al.* 2005. Body weight supported treadmill training in acute spinal cord injury: impact on muscle and bone. *Spinal Cord* **43:** 649–657.

69. Alekna, V., M. Tamulaitiene, T. Sinevicius & A. Juocevicius. 2008. Effect of weight-bearing activities on bone mineral

density in spinal cord injured patients during the period of the first two years. *Spinal Cord* **46:** 727–732.

70. Lofvenmark, I., L. Werhagen & C. Norrbrink. 2009. Spasticity and bone density after a spinal cord injury. *J. Rehabil. Med.* **41:** 1080–1084.

71. Eser, P., A. Frotzler, Y. Zehnder, *et al.* 2005. Assessment of anthropometric, systemic, and lifestyle factors influencing bone status in the legs of spinal cord injured individuals. *Osteoporos. Int.* **16:** 26–34.

72. Gorgey, A.S. & G.A. Dudley. 2008. Spasticity may defend skeletal muscle size and composition after incomplete spinal cord injury. *Spinal Cord* **46:** 96–102.

73. Zaidi, M. 2007. Skeletal remodeling in health and disease. *Nat. Med.* **13:** 791–801.

74. Pavalko, F.M. *et al.* 2003. A model for mechanotransduction in bone cells: the load-bearing mechanosomes. *J. Cell Biochem.* **88:** 104–112.

75. Han, Y., S.C. Cowin, M.B. Schaffler & S. Weinbaum. 2004. Mechanotransduction and strain amplification in osteocyte cell processes. *Proc. Natl. Acad. Sci. USA* **101:** 16689–16694.

76. Jiang, S.D., L.S. Jiang & L.Y. Dai. 2006. Mechanisms of osteoporosis in spinal cord injury. *Clin. Endocrinol. (Oxf)* **65:** 555–565.

77. Bonewald, L.F. & M.L. Johnson. 2008. Osteocytes, mechanosensing and Wnt signaling. *Bone* **42:** 606–615.

78. McBeath, R., D.M. Pirone, C.M. Nelson, *et al.* 2004. Cell shape, cytoskeletal tension, and RhoA regulate stem cell lineage commitment. *Dev. Cell* **6:** 483–495.

79. Bergmann, P., A. Heilporn, A. Schoutens. 1977. Longitudinal study of calcium and bone metabolism in paraplegic patients. *Paraplegia* **15:** 147–159.

80. Roberts, D. *et al.* Longitudinal study of 1998. bone turnover after acute spinal cord injury. *J. Clin. Endocrinol. Metab.* **83:** 415–422.

81. Pietschmann, P. *et al.* 1992. Increased serum osteocalcin levels in patients with paraplegia. *Paraplegia* **30:** 204–209.

82. Uebelhart, D. *et al.* 1994. Early modifications of biochemical markers of bone metabolism in spinal cord injury patients. A preliminary study. *Scand. J. Rehabil. Med.* **26:** 197–202.

83. Reiter, A.L., A. Volk, J. Vollmar, *et al.* 2007. Changes of basic bone turnover parameters in short-term and long-term patients with spinal cord injury. *Eur. Spine J.* **16:** 771–776.

84. Morse, L.R. *et al.* 2008. Age and motor score predict osteoprotegerin level in chronic spinal cord injury. *J. Musculoskelet Neuronal Interact* **8:** 50–57.

85. Demulder, A. *et al.* 1998. Increased osteoclast-like cells formation in long-term bone marrow cultures from patients with a spinal cord injury. *Calcif. Tissue Int.* **63:** 396–400.

86. Sugawara, H., T.A. Linsenmeyer, H. Beam, & J.R. Parsons. 1998. Mechanical properties of bone in a paraplegic rat model. *J. Spinal Cord Med.* **21:** 302–308.

87. Morse, L. *et al.* 2008. Spinal cord injury causes rapid osteoclastic resorption and growth plate abnormalities in growing rats (SCI-induced bone loss in growing rats). *Osteoporos Int.* **19:** 645–652.

88. Jiang, S.D., L.S. Jiang & L.Y. Dai. 2007. Effects of spinal cord injury on osteoblastogenesis, osteoclastogenesis and gene expression profiling in osteoblasts in young rats. *Osteoporos Int.* **18,** 339–349.

89. Jiang, S.D., L.S. Jiang & L.Y. Dai. 2007. Changes in bone mass, bone structure, bone biomechanical properties, and bone metabolism after spinal cord injury: a 6-month longitudinal study in growing rats. *Calcif. Tissue Int.* **80:** 167–175.

90. Picard, S., N.P. Lapointe, J.P. Brown & P.A. Guertin. 2008. Histomorphometric and densitometric changes in the femora of spinal cord transected mice. *Anat. Rec. (Hoboken)* **291:** 303–307.

91. Bracken, M.B. & T.R. Holford. 2002. Neurological and functional status 1 year after acute spinal cord injury: estimates of functional recovery in National Acute Spinal Cord Injury Study II from results modeled in National Acute Spinal Cord Injury Study III. *J. Neurosurg.* **96:** 259–266.

92. Bracken, M.B. *et al.* 1998. Methylprednisolone or tirilazad mesylate administration after acute spinal cord injury: 1-year follow up. Results of the third National Acute Spinal Cord Injury randomized controlled trial. *J. Neurosurg.* **89:** 699–706.

93. Bracken, M.B. *et al.* 1997. Administration of methylprednisolone for 24 or 48 hours or tirilazad mesylate for 48 hours in the treatment of acute spinal cord injury. Results of the Third National Acute Spinal Cord Injury Randomized Controlled Trial. National Acute Spinal Cord Injury Study. *JAMA* **277:** 1597–1604.

94. Young, W. & M.B. Bracken. 1992. The Second National Acute Spinal Cord Injury Study. *J. Neurotrauma* **9**(Suppl 1): S397–405.

95. Bracken, M.B. *et al.* 1992. Methylprednisolone or naloxone treatment after acute spinal cord injury: 1-year follow-up data. Results of the second National Acute Spinal Cord Injury Study. *J. Neurosurg.* **76:** 23–31.

96. Bracken, M.B. 1991. Treatment of acute spinal cord injury with methylprednisolone: results of a multicenter, randomized clinical trial. *J. Neurotrauma* **8**(Suppl 1): S47–50; discussion S51–S52.

97. Bracken, M.B. *et al.* 1990. A randomized, controlled trial of methylprednisolone or naloxone in the treatment of acute spinal-cord injury. Results of the Second National Acute Spinal Cord Injury Study. *N. Engl. J. Med.* **322:** 1405–1411.

98. Qian, T. *et al.* 2005. High-dose methylprednisolone may cause myopathy in acute spinal cord injury patients. *Spinal Cord* **43:** 199–203.

99. Nicholas, J.J., D.H. Streeten & L. Jivoff. 1969. A study of pituitary and adrenal function in patients with traumatic injuries of the spinal cord. *J. Chronic. Dis.* **22:** 463–471.

100. Zeitzer, J.M., N.T. Ayas, S.A. Shea, *et al.* 2000. Absence of detectable melatonin and preservation of cortisol and thyrotropin rhythms in tetraplegia. *J. Clin. Endocrinol. Metab.* **85:** 2189–2196.

101. Dalla Libera, L. *et al.* 2001. Beneficial effects on skeletal muscle of the angiotensin II type 1 receptor blocker irbesartan in experimental heart failure. *Circulation* **103:** 2195–2200.

102. Brink, M. *et al.* 2001. Angiotensin II induces skeletal muscle wasting through enhanced protein degradation and

down-regulates autocrine insulin-like growth factor I. *Endocrinology*. **142:** 1489–1496.

103. Wecht, J.M. *et al.* 2009. Orthostatic responses to nitric oxide synthase inhibition in persons with tetraplegia. *Arch. Phys. Med. Rehabil.* **90:** 1428–1434.

104. Bhasin, S. *et al.* 2001. Testosterone dose-response relationships in healthy young men. *Am. J. Physiol. Endocrinol. Metab.* **281:** E1172–E1181.

105. Bhasin, S. *et al.* 2005. Older men are as responsive as young men to the anabolic effects of graded doses of testosterone on the skeletal muscle. *J. Clin. Endocrinol. Metab.* **90:** 678–688.

106. Mauras, N. *et al.* 1998. Testosterone deficiency in young men: marked alterations in whole body protein kinetics, strength, and adiposity. *J. Clin. Endocrinol. Metab.* **83:** 1886–1892.

107. Kostovski, E., P.O. Iversen, K. Birkeland, *et al.* 2008. Decreased levels of testosterone and gonadotrophins in men with long-standing tetraplegia. *Spinal Cord.* **46:** 559–564.

108. Strawford, A. *et al.* 1999. Resistance exercise and supra-physiologic androgen therapy in eugonadal men with HIV-related weight loss: a randomized controlled trial. *JAMA.* **281:** 1282–1290.

109. Earthman, C.P., P.M. Reid, I.T. Harper, *et al.* 2002. Body cell mass repletion and improved quality of life in HIV-infected individuals receiving oxandrolone. *JPEN J. Parenter. Enteral. Nutr.* **26:** 357–365.

110. Montano, M. *et al.* 2007. Transcriptional profiling of testosterone-regulated genes in the skeletal muscle of human immunodeficiency virus-infected men experiencing weight loss. *J. Clin. Endocrinol. Metab.* **92:** 2793–2802.

111. Schols, A.M., P.B. Soeters, R. Mostert, *et al.* 1995. Physiologic effects of nutritional support and anabolic steroids in patients with chronic obstructive pulmonary disease. A placebo-controlled randomized trial. *Am. J. Respir. Crit. Care Med.* **152:** 1268–1274.

112. Yeh, S.S., B. DeGuzman & T. Kramer. 2002. Reversal of COPD-associated weight loss using the anabolic agent oxandrolone. *Chest.* **122:** 421–428.

113. Demling, R.H. & L. DeSanti. 1997. Oxandrolone, an anabolic steroid, significantly increases the rate of weight gain in the recovery phase after major burns. *J. Trauma* **43:** 47–51.

114. Wolf, S.E. *et al.* 2006. Effects of oxandrolone on outcome measures in the severely burned: a multicenter prospective randomized double-blind trial. *J. Burn Care Res.* **27:** 131–139. discussion 140–131.

115. Crawford, B.A., P.Y. Liu, M.T. Kean, *et al.* 2003. Randomized placebo-controlled trial of androgen effects on muscle and bone in men requiring long-term systemic glucocorticoid treatment. *J. Clin. Endocrinol. Metab.* **88:** 3167–3176.

116. Wimalawansa, S.M. *et al.* 1999. Reversal of weightlessness-induced musculoskeletal losses with androgens: quantification by MRI. *J. Appl. Physiol.* **86:** 1841–1846.

117. Taylor, D.C., D.E. Brooks & J.B. Ryan. 1999. Anabolic-androgenic steroid administration causes hypertrophy of immobilized and nonimmobilized skeletal muscle in a sedentary rabbit model. *Am. J. Sports Med.* **27:** 718–727.

118. Zhao, W. *et al.* 2008. Testosterone protects against dexamethasone-induced muscle atrophy, protein degradation and MAFbx upregulation. *J. Steroid Biochem. Mol. Biol.* **110:** 125–129.

119. Gregory, C.M., K. Vandenborne, H.F. Huang, *et al.* 2003. Effects of testosterone replacement therapy on skeletal muscle after spinal cord injury. *Spinal Cord.* **41:** 23–28.

120. Sheffield-Moore, M. *et al.* 1999. Short-term oxandrolone administration stimulates net muscle protein synthesis in young men. *J. Clin. Endocrinol. Metab.* **84:** 2705–2711.

121. Sarbassov, D.D., S.M. Ali & D.M. Sabatini. 2005. Growing roles for the mTOR pathway. *Curr. Opin. ell Biol.* **17:** 596–603.

122. Xu, T. *et al.* 2004. Phosphorylation of p70s6 kinase is implicated in androgen-induced levator ani muscle anabolism in castrated rats. *J. Steroid Biochem. Mol. Biol.* **92:** 447–454.

123. Rommel, C. *et al.* 2001. Mediation of IGF-1-induced skeletal myotube hypertrophy by PI(3)K/Akt/mTOR and PI(3)K/Akt/GSK3 pathways. *Nat. Cell Biol.* **3:** 1009–1013.

124. Schakman, O. *et al.* 2005. Insulin-like growth factor-I gene transfer by electroporation prevents skeletal muscle atrophy in glucocorticoid-treated rats. *Endocrinology* **146:** 1789–1797.

125. Dehoux, M. *et al.* 2004. Role of the insulin-like growth factor I decline in the induction of atrogin-1/MAFbx during fasting and diabetes. *Endocrinology* **145:** 4806–4812.

126. Alzghoul, M.B., D. Gerrard, B.A. Watkins & K. Hannon. 2004. Ectopic expression of IGF-I and Shh by skeletal muscle inhibits disuse-mediated skeletal muscle atrophy and bone osteopenia in vivo. *FASEB J.* **18:** 221–223.

127. Wu, Y. *et al.* 2007. Identification of Androgen Response Elements in the IGF-1 Upstream Promoter. *Endocrinology* **148:** 2984–2993.

128. Ferrando, A.A. *et al.* 2002. Testosterone administration to older men improves muscle function: molecular and physiological mechanisms. *Am. J. Physiol. Endocrinol. Metab.* **282:** E601–607.

129. Sofer, A., K. Lei, C.M. Johannessen, & L.W. Ellisen, 2005. Regulation of mTOR and cell growth in response to energy stress by REDD1. *Mol. Cell Biol.* **25:** 5834–5845.

130. Wang, H., N. Kubica, L.W. Ellisen, L.S. Jefferson, & S.R. Kimball. 2006. Dexamethasone represses signaling through the mammalian target of rapamycin in muscle cells by enhancing expression of REDD1. *J. Biol. Chem.* **281:** 39128–39134.

131. Wu, Y. *et al.* 2009. REDD1 Is a Major Target of Testosterone Action in Preventing Dexamethasone-Induced Muscle Loss. *Endocrinology*.

132. Ferrando, A.A., M. Sheffield-Moore, S.E. Wolf, *et al.* 2001. Testosterone administration in severe burns ameliorates muscle catabolism. *Crit. Care Med.* **29:** 1936–1942.

133. Sandri, M. *et al.* 2004. Foxo transcription factors induce the atrophy-related ubiquitin ligase atrogin-1 and cause skeletal muscle atrophy. *Cell* **117:** 399–412.

134. Stitt, T.N. *et al.* 2004. The IGF-1/PI3K/Akt pathway prevents expression of muscle atrophy-induced ubiquitin ligases by inhibiting FOXO transcription factors. *Mol. Cell* **14:** 395–403.

135. Bauman, W.A. *et al.* 2006. Effect of low-dose baclofen administration on plasma insulin-like growth factor-I in persons with spinal cord injury. *J. Clin. Pharmacol.* **46:** 476–482.

136. Hawley, J.A. 2004. Exercise as a therapeutic intervention for the prevention and treatment of insulin resistance. *Diabetes Metab. Res. Rev.* **20:** 383–393.

137. Frosig, C. & E.A. Richter. 2009. Improved insulin sensitivity after exercise: focus on insulin signaling. *Obesity (Silver Spring)* **17**(Suppl 3): S15–S20.

138. Mohr, T. *et al.* 2001. Insulin action and long-term electrically induced training in individuals with spinal cord injuries. *Med. Sci. Sports Exerc.* **33:** 1247–1252.

139. Spungen, A.M., J. Wang, R.N. Pierson, *et al.* 2000. Soft tissue body composition differences in monozygotic twins discordant for spinal cord injury. *J. Appl. Physiol.* **88:** 1310–1315.

140. Spungen, A.M. *et al.* 2003. Factors influencing body composition in persons with spinal cord injury: a cross-sectional study. *J. Appl. Physiol.* **95:** 2398–2407.

141. Elder, C.P., D.F. Apple, C.S. Bickel, *et al.* 2004. Intramuscular fat and glucose tolerance after spinal cord injury—a cross-sectional study. *Spinal Cord* **42:** 711–716.

142. Gorgey, A.S. & G.A. Dudley. 2007. Skeletal muscle atrophy and increased intramuscular fat after incomplete spinal cord injury. *Spinal Cord* **45:** 304–309.

143. Goodpaster, B.H., F.L. Thaete, J.A. Simoneau & D.E. Kelley. 1997. Subcutaneous abdominal fat and thigh muscle composition predict insulin sensitivity independently of visceral fat. *Diabetes* **46:** 1579–1585.

144. Goodpaster, B.H., J. He, S. Watkins & D.E. Kelley. 2001. Skeletal muscle lipid content and insulin resistance: evidence for a paradox in endurance-trained athletes. *J. Clin. Endocrinol. Metab.* **86:** 5755–5761.

145. Hafer-Macko, C.E., S. Yu, A.S. Ryan, *et al.* 2005. Elevated tumor necrosis factor-alpha in skeletal muscle after stroke. *Stroke* **36:** 2021–2023.

146. Szulc, P. *et al.* 2001. Bioavailable estradiol may be an important determinant of osteoporosis in men: the MINOS study. *J. Clin. Endocrinol. Metab.* **86:** 192–199.

147. Khosla, S., L.J. Melton, 3rd, E.J. Atkinson & W.M. O'Fallon 2001. Relationship of serum sex steroid levels to longitudinal changes in bone density in young versus elderly men. *J. Clin. Endocrinol. Metab.* **86:** 3555–3561.

148. Riggs, B.L., S. Khosla & L.J. Melton, 3rd. 1998. A unitary model for involutional osteoporosis: estrogen deficiency causes both type I and type II osteoporosis in postmenopausal women and contributes to bone loss in aging men. *J. Bone Miner. Res.* **13:** 763–773.

149. Khosla, S., L.J. Melton, 3rd & B.L. Riggs. 2002. Clinical review 144: Estrogen and the male skeleton. *J. Clin. Endocrinol. Metab.* **87:** 1443–1450.

150. Gennari, L., S. Khosla & J.P. Bilezikian. 2008. Estrogen and fracture risk in men. *J. Bone. Miner. Res.* **23:** 1548–1551.

151. Vandenput,L. & C. Ohlsson. 2009. Estrogens as regulators of bone health in men. *Nat. Rev. Endocrinol.* **5:** 437–443.

152. Kley, H.K., T. Deselaers, H. Peerenboom & H.L. Kruskemper. 1980. Enhanced conversion of androstenedione to estrogens in obese males. *J. Clin. Endocrinol. Metab.* **51:** 1128–1132.

153. Simpson, E.R. 2000. Role of aromatase in sex steroid action. *J. Mol. Endocrinol.* **25:** 149–156.

154. Napoli, N. *et al.* 2007. Estrogen metabolism modulates bone density in men. *Calcif. Tissue Int.* **80:** 227–232.

155. LeBlanc, E.S. *et al.* 2009. The effects of serum testosterone, estradiol, and sex hormone binding globulin levels on fracture risk in older men. *J. Clin. Endocrinol. Metab.* **94:** 3337–3346.

156. Gu, G., T.A. Hentunen, M. Nars, *et al.* 2005. Estrogen protects primary osteocytes against glucocorticoid-induced apoptosis. *Apoptosis* **10:** 583–595.

157. Huang, Q.Y., G.H. Li & A.W. Kung. 2009. The -9247 T/C polymorphism in the SOST upstream regulatory region that potentially affects C/EBPalpha and FOXA1 binding is associated with osteoporosis. *Bone* **45:** 289–294.

158. Notelovitz, M. 2002. Androgen effects on bone and muscle. *Fertil Steril* **77**(Suppl 4), S34–S41.

159. Francis, R.M.N. 1999. The effects of testosterone on osteoporosis in men. *Clin. Endocrinol. (Oxf)* **50:** 411–414.

160. Anderson, F.H., R.M. Francis, P.L. Selby & C. Cooper. 1998. Sex hormones and osteoporosis in men. *Calcif. Tissue Int.* **62:** 185–188.

161. Pederson, L. *et al.* 1999. Androgens regulate bone resorption activity of isolated osteoclasts in vitro. *Proc. Natl. Acad. Sci. USA* **96:** 505–510.

162. Cardozo, C.P. *et al.* 2010. Nandrolone slows hindlimb bone loss in a rat model of bone loss due to denervation. *Ann. N. Y. Acad. Sci.* **1192:** 303–306.

163. Bauman, W.A. & A.M. Spungen. 2000. Metabolic changes in persons after spinal cord injury. *Phys. Med. Rehabil. Clin. N. Am.* **11:** 109–140.

164. Claus-Walker, J., R.J. Campos, R.E. Carter, *et al.* 1972. Calcium excretion in quadriplegia. *Arch. Phys. Med. Rehabil.* **53:** 14–20.

165. Maynard, F.M. 1986. Immobilization hypercalcemia following spinal cord injury. *Arch. Phys. Med. Rehabil.* **67:** 41–44.

166. Naftchi, N.E., A.T. Viau, G.H. Sell & E.W. Lowman. 1980. Mineral metabolism in spinal cord injury. *Arch. Phys. Med. Rehabil.* **61:** 139–142.

167. Tori, J.A. & L.L. Hil. 1978. Hypercalcemia in children with spinal cord injury. *Arch. Phys. Med. Rehabil.* **59:** 443–446.

168. Kaplan, P.E., B. Gandhavadi, L. Richards & J. Goldschmidt. 1978. Calcium balance in paraplegic patients: influence of injury duration and ambulation. *Arch. Phys. Med. Rehabil.* **59:** 447–450.

169. Stewart, A.F., M. Adler, C.M. Byers, *et al.* 1982. Calcium homeostasis in immobilization: an example of resorptive hypercalciuria. *N. Engl. J. Med.* **306:** 1136–1140.

170. Mechanick, J.I. *et al.* 1997. Parathyroid hormone suppression in spinal cord injury patients is associated with the degree of neurologic impairment and not the level of injury. *Arch. Phys. Med. Rehabil.* **78:** 692–696.

171. Loomis, W.F. 1967. Skin-pigment regulation of vitamin-D biosynthesis in man. *Science* **157:** 501–506.

172. Comarr, A.E., R.H. Hutchinson & E. Bors. 1962. Extremity fractures of patients with spinal cord injuries. *Am. J. Surg.* **103:** 732–739.

173. Hahn, T.J., B.A. Hendin, C.R. Scharp & J.G. Haddad Jr. 1972. Effect of chronic anticonvulsant therapy on serum 25–hydroxycalciferol levels in adults. *N. Engl. J. Med.* **287:** 900–904.

174. Bauman, W.A., Y.G. Zhong & E. Schwartz. 1995. Vitamin D deficiency in veterans with chronic spinal cord injury. *Metabolism* **44:** 1612–1616.

175. Miao, D. *et al.* 2004. Skeletal abnormalities in Pth-null mice are influenced by dietary calcium. *Endocrinology* **145:** 2046–2053.

176. Lips, P., W.H. Hackeng, M.J. Jongen, *et al.* 1983. Seasonal variation in serum concentrations of parathyroid hormone in elderly people. *J. Clin. Endocrinol. Metab.* **57:** 204–206.

177. Chapuy, M.C. *et al.* 1983. Healthy elderly French women living at home have secondary hyperparathyroidism and high bone turnover in winter. EPIDOS Study Group. *J. Clin. Endocrinol. Metab.* **81:** 1129–1133.

178. Bauman, W.A., R.L. Zhang, N. Morrison & A.M. Spungen. 2009. Acute suppression of bone turnover with calcium infusion in persons with spinal cord injury. *J. Spinal Cord Med.* **32:** 398–403.

179. Scelsi, R. *et al.* 1986. Skeletal muscle changes following myelotomy in paraplegic patients. *Paraplegia* **24:** 250–259.

180. Edgerton, V.R. *et al.* 2008. Training locomotor networks. *Brain Res. Rev.* **57:** 241–254.

181. Hyatt, J.P., R.R. Roy, K.M. Baldwin & V.R. Edgerton. 2003. Nerve activity-independent regulation of skeletal muscle atrophy: role of MyoD and myogenin in satellite cells and myonuclei. *Am. J. Physiol. Cell Physiol.* **285:** C1161–C1173.

182. Hyatt, J.P., R.R. Roy, K.M. Baldwin, *et al.* 2006. Activity-unrelated neural control of myogenic factors in a slow muscle. *Muscle Nerve* **33:** 49–60.

183. Takeda, S. *et al.* 2002. Leptin regulates bone formation via the sympathetic nervous system. *Cell* **111:** 305–317.

184. Sato, S. *et al.* 2007. Central control of bone remodeling by neuromedin U. *Nat. Med.* **13:** 1234–1240.

185. Bjurholm, A., A. Kreicbergs, E. Brodin & M. Schultzberg. 1988. Substance P- and CGRP-immunoreactive nerves in bone. *Peptides* **9:** 165–171.

186. Hukkanen, M. *et al.* 1992. Innervation of bone from healthy and arthritic rats by substance P and calcitonin gene related peptide containing sensory fibers. *J. Rheumatol.* **19:** 1252–1259.

187. Juarranz, Y. *et al.* 2005. Protective effect of vasoactive intestinal peptide on bone destruction in the collagen-induced arthritis model of rheumatoid arthritis. *Arthritis Res. Ther.* **7:** R1034–R1045.

188. Mukohyama, H., M. Ransjo, H. Taniguchi, *et al.* 2000. The inhibitory effects of vasoactive intestinal peptide and pituitary adenylate cyclase-activating polypeptide on osteoclast formation are associated with upregulation of osteoprotegerin and downregulation of RANKL and RANK. *Biochem. Biophys. Res. Commun.* **271:** 158–163.

189. Fu, L., M. S. Patel, A. Bradley, *et al.* 2005. The molecular clock mediates leptin-regulated bone formation. *Cell* **122:** 803–815.

190. Olive, J.L., G.A. Dudley & K.K. McCully. 2003. Vascular remodeling after spinal cord injury. *Med. Sci. Sports Exerc.* **35:** 901–907.

191. Castro, M.J., D.F. Apple Jr., S. Rogers & G.A. Dudley. 2000. Influence of complete spinal cord injury on skeletal muscle mechanics within the first 6 months of injury. *Eur. J. Appl. Physiol.* **81:** 128–131.

192. Wilmet, E., A.A. Ismail, A. Heilporn, *et al.* 1995. Longitudinal study of the bone mineral content and of soft tissue composition after spinal cord section. *Paraplegia* **33:** 674–677.

193. McDonald, C.M., A.L. Abresch-Meyer, M.D. Nelson & L.M. Widman. 2007. Body mass index and body composition measures by dual x-ray absorptiometry in patients aged 10 to 21 years with spinal cord injury. *J. Spinal Cord Med.* **30**(Suppl 1): S97–S104.

194. Trudel, G. *et al.* 2009. Bone marrow fat accumulation after 60 days of bed rest persisted 1 year after activities were resumed along with hemopoietic stimulation: the Women International Space Simulation for Exploration study. *J. Appl. Physiol.* **107:** 540–548.

195. Minaire, P., C. Edouard, M. Arlot & P.J. Meunier. 1984. Marrow changes in paraplegic patients. *Calcif. Tissue Int.* **36:** 338–340.

196. Jee, W.S., T.J. Wronski, E.R. Morey & D.B. Kimmel. 1983. Effects of spaceflight on trabecular bone in rats. *Am. J. Physiol.* **244:** R310–R314.

197. Wronski, T.J. & E.R. Morey. 1982. Skeletal abnormalities in rats induced by simulated weightlessness. *Metab. Bone Dis. Relat. Res.* **4:** 69–75.

198. Schopp, L.H. *et al.* 2006. Testosterone levels among men with spinal cord injury admitted to inpatient rehabilitation. *Am. J. Phys. Med. Rehabil.* **85:** 678–684; quiz 685–677.

199. Singh, R. *et al.* 2006. Testosterone inhibits adipogenic differentiation in 3T3-L1 cells: nuclear translocation of androgen receptor complex with beta-catenin and T-cell factor 4 may bypass canonical Wnt signaling to down-regulate adipogenic transcription factors. *Endocrinology* **147:** 141–154.

200. Menagh, P. *et al.* 2009. Growth hormone regulates the balance between bone formation and bone marrow adiposity. *J. Bone Miner. Res.*

201. Johnson, T.E., R. Vogel, S. J. Rutledge, *et al.* 1999. Thiazolidinedione effects on glucocorticoid receptor-mediated gene transcription and differentiation in osteoblastic cells. *Endocrinology* **140:** 3245–3254.

202. Ishida, Y. & J.N. Heersche. 1998. Glucocorticoid-induced osteoporosis: both in vivo and in vitro concentrations of glucocorticoids higher than physiological levels attenuate osteoblast differentiation. *J. Bone Miner. Res.* **13:** 1822–1826.

203. David, V. *et al.* 2007. Mechanical loading down-regulates peroxisome proliferator-activated receptor gamma in bone marrow stromal cells and favors osteoblastogenesis at the expense of adipogenesis. *Endocrinology* **148:** 2553–2562.

204. Zalavras, C., S. Shah, M.J. Birnbaum & B. Frenkel. 2003. Role of apoptosis in glucocorticoid-induced osteoporosis

and osteonecrosis. *Crit. Rev. Eukaryot. Gene Expr.* **13:** 221–235.

205. Karastergiou, K. & V. Mohamed-Ali. The autocrine and paracrine roles of adipokines. *Mol. Cell Endocrinol.* **318:** 69–78.

206. Dyck, D.J. 2009. Adipokines as regulators of muscle metabolism and insulin sensitivity. *Appl. Physiol. Nutr. Metab.* **34:** 396–402.

207. Lago, F., R. Gomez, J.J. Gomez-Reino, C. Dieguez & O.

Gualillo. 2009. Adipokines as novel modulators of lipid metabolism. *Trends Biochem. Sci.* **34:** 500–510.

208. Bauman, W.A., A.M. Spungen, Y.G. Zhong & C.V. Mobbs. 1996. Plasma leptin is directly related to body adiposity in subjects with spinal cord injury. *Horm. Metab. Res.* **28:** 732–736.

209. Maruyama, Y. *et al.* 2008. Serum leptin, abdominal obesity and the metabolic syndrome in individuals with chronic spinal cord injury. *Spinal Cord* **46:** 494–499.

Ann. N.Y. Acad. Sci. ISSN 0077-8923

ANNALS OF THE NEW YORK ACADEMY OF SCIENCES
Issue: *Molecular and Integrative Physiology of the Musculoskeletal System*

# Bone physiology and therapeutics in chronic critical illness

Michael A. Via,[1] Emily Jane Gallagher,[2] and Jeffrey I. Mechanick[2]

[1]Division of Endocrinology and Metabolism, Beth Israel Medical Center, Albert Einstein College of Medicine, Bronx, New York. [2]Division of Endocrinology, Diabetes and Bone Disease, Mount Sinai School of Medicine, New York, New York

Address for correspondence: Michael A. Via, M.D., Division of Endocrinology and Metabolism, Beth Israel Medical Center, Albert Einstein College of Medicine, 55 East 34th St., New York, NY 10016. mvia@chpnet.org

Modern medical practices allow patients to survive acute insults and be sustained by machinery and medicines for extended periods of time. We define chronic critical illness as a later stage of prolonged critical illness that requires tracheotomy. These patients have persistent elevations of inflammatory cytokines, diminished hypothalamic–pituitary function, hypercatabolism, immobilization, and malnutrition. The measurement of bone turnover markers reveals markedly enhanced osteoclastic bone resorption that is uncoupled from osteoblastic bone formation. We review the mechanisms by which these factors contribute to the metabolic bone disease of chronic critical illness and suggest potential therapeutics.

Keywords: metabolic bone disease; chronic critical illness; allostasis; bone resorption

## Introduction

In 1952, medical intensive care, as we know it today, emerged amidst the polio epidemic in Europe. Despite much skepticism, Bjørn Ibsen, a Danish anesthesiologist, successfully ventilated a 12-year-old girl with bulbar poliomyelitis, and subsequently many other patients with respiratory failure from polio were supported by artificial ventilation, which reduced the mortality rate from 87% to less than 15%. He later went on to create the world's first intensive care unit (ICU) in Copenhagen.[1] Critical illness is generally defined as a life-threatening condition that requires continuous monitoring and comprehensive care.[2] Throughout human history before the emergence of critical care medicine, a person suffering a calamitous insult from infection or trauma would launch his or her cytokine and hormone defenses, ultimately overcoming the insult and surviving, or alternatively, succumbing to the insult and dying. This process of physiological adaptation to maintain biological stability (homeostasis) is known as allostasis, and is mediated through inflammatory cytokine release and set-point changes of the neuroendocrine system (immune–neuroendocrine axis [INA] modulation).

In the modern era of ventilators, pressors, advanced-generation antibiotics, renal replacement therapy, and organ transplantation, the rapid demise of ICU patients is no longer inevitable. On the contrary, modern technological support does not guarantee rapid recuperation, and individuals may transition from an acute critical illness (ACI) stage, through a prolonged ACI (PACI) stage, to a chronic critical illness (CCI) stage, from which they may or may not enter a recovery from critical illness stage. The allostatic processes that prioritize survival during ACI are exposed to Darwinian evolutionary pressures of acute stress but not to evolutionary pressures of PACI–CCI because they do not exist. Therefore, non-Darwinian allostatic processes of CCI may be detrimental to recovery. In this context of CCI, we will focus on one pathophysiological state, that is, metabolic bone disease.

## Chronic critical illness

Failure to recover from ACI is generally ascribed to multi-organ failure or infectious complications, and PACI begins after 3–5 days of ACI in those individuals who remain critically ill.[3–5] ACI and PACI are distinguished by distinct neuroendocrine responses,

doi: 10.1111/j.1749-6632.2010.05807.x

involving the hypothalamic–pituitary axes governing growth hormone (GH) and insulin-like growth factor-1 (IGF-1), cortisol, thyroid hormone, and gonadal hormones.[6–8] If recovery from PACI does not occur and the tracheotomy is performed, the patient enters the CCI stage.

The term "chronically critically ill" was coined by Girard and Raffin[9] to describe ICU patients who do not survive despite extraordinary life support for weeks to months. Others have defined individuals with CCI as those who survive critical illness but require long-term ventilation and intensive nursing care after receiving therapy for the primary disease, or those who because of underlying illness or complications suffer a prolonged and complicated ICU course.[9–13] For the purposes of research, it is often defined from the time of tracheostomy placement, as this is an easily defined event in the patient's hospitalization. From the onset of the ACI, tracheostomy placement usually occurs after 10–14 days and generally signifies an expectation that the patient is unlikely to immediately recover and unlikely to immediately die, but rather is likely have a protracted ventilator-dependent course.[14] The metabolic, neuroendocrine, neuropsychiatric, and immunological derangements associated with failure to be liberated from mechanical ventilation define the "CCI syndrome."[15]

In view of the aging population, the increased burden of chronic illness in the population, and the exceptional ability of medical professionals to support patients during ACI, the annual number of CCI patients is expected to more than double in the United States, from 250,000 cases in the year 2000 to over 600,000 cases by 2020.[16]

## Bone disease in CCI

CCI is associated with a high risk of disability and distress. Patients that develop CCI have greater than 50% mortality six months after hospital discharge.[15] There are many psychological and physical consequences of CCI, one of which is metabolic bone disease. The metabolic bone disease associated with CCI is associated with bone hyperresorption, vitamin D deficiency and secondary hyperparathyroidism, prolonged immobilization, systemic inflammation, and certain medications such as glucocorticoids.[17] These patients are at high risk of vitamin D deficiency from poor nutrition, lack of sunlight, renal disease, and hepatic disease.[14]

Prolonged immobilization leads to increased calcium resorption from bones and hypercalciuria, inhibiting parathyroid hormone (PTH) and 1,25-dihydroxy vitamin D (1,25-D) formation.[18] These abnormalities were demonstrated in a study of 49 patients with CCI, in which only four patients (9%) had normal 24 h urine N-telopeptide of type 1 collagen (NTX) levels, whereas the remainder had evidence of bone hyperresorption with elevated NTX levels. Forty-two percent of these patients demonstrated a predominant vitamin D deficiency, 9% had evidence of hyperresorption related to prolonged immobilization, and 49% had evidence of overlap of metabolic bone disease from both vitamin D deficiency and immobilization.[14] Markers of accelerated bone turnover were seen to correlate with immobility, hip fracture, and all-cause mortality, after adjustment for other risk factors, in a prospective study of 1112 individuals living in residential care.[19] This suggests that markers of increased bone resorption may not only merely be a predictor of future fracture risk but may also correlate with mortality in an elderly immobilized population such as those with CCI.

## Osteoblasts and osteoclasts

Bone remodeling occurs on a continuous basis due to the coordinated activity of osteoclasts that resorb bone and osteoblasts that lay down new bone. Markers of bone loss or osteoclast function include C-terminal telopeptide of type 1 collagen (CTX) and NTX. Markers of osteoblast function are N-terminal propeptide of type 1 collagen (P1NP), C-terminal propeptide of type 1 collagen (PICP), bone-specific alkaline phosphatase (sALP), and osteocalcin (OCN). These biomarkers have been used to study the process of bone remodeling in CCI patients.

Osteoblasts originate from multipotent mesenchymal stem cells that have the potential to differentiate into osteoblasts, adipocytes, chondrocytes, myocytes, or fibroblasts. Osteoblasts produce osteoid, which is subsequently mineralized to form new bone. In addition, osteoblasts produce a variety of GHs, including IGF-1, transforming growth factor-beta, and basic fibroblast growth factor, among others. Osteoblasts have cell surface receptors for PTH, PTH-related protein, thyroid hormone, GH, insulin, progesterone, and prolactin as well as nuclear receptors for estrogens, androgens, vitamin D, and retinoids.[20] Osteoclasts are derived from the hematopoietic mononuclear cell lineage. The function of osteoclasts is regulated by cytokines and systemic hormones, including receptor activator for nuclear factor kappa-B (NF-κB) ligand (RANKL), calcitonin, androgens, thyroid hormone, insulin, PTH, IGF-1, interleukin (IL)-1, colony-stimulating factor-1, platelet-derived growth factor, follicle-stimulating hormone, and thyroid-stimulating hormone.[20,21] The process of bone remodeling involves the coordinated coupling of osteoclast and osteoblast activity, but this balance is affected by cytokine-mediated inflammation and neuroendocrine axis dampening that is characteristic of CCI.[22]

Osteoclast function is regulated via RANKL and osteoprotegerin (OPG) that are produced by osteoblasts. Binding of OPG to RANKL prevents osteoclast differentiation and promotes osteoclast apoptosis.[23] RANK recruits adaptor proteins, such as the tumor necrosis factor (TNF)-receptor-associated factor 6 (TRAF6), with resultant activation of NF-κB and mitogen-activated protein kinase (MAPK), leading to the expression of the target genes, activator protein-1 (AP-1), and nuclear factor of activated T cells and cytoplasmic calcineurin-dependent 1. Activation of these genes allows for fusion of pre-osteoclasts and differentiation.[24,25]

Osteoblast differentiation and bone formation occurs by the activation of genes as a result of signaling mediated by the osteoblast tyrosine kinase transmembrane receptor, EphB4. Through a bidirectional signal pathway, EphB4 induces its ligand, osteoclast cell surface protein ephrinB2, which inhibits osteoclast differentiation.[26] These processes allow for coupling of bone resorption and formation by osteoblast–osteoclast communication.

## Inflammation and bone remodeling

The immune system has the ability to affect the balance of bone resorption and formation by inducing osteoclast differentiation. In addition to osteoblasts, RANKL is also produced by monocytes, neutrophils, and lymphocytes, which leads to activation of osteoclasts. In addition, these cells produce inflammatory cytokines, such as TNF-α, and ILs (e.g., IL-1, IL-3, IL-6, and IL-17), which may increase osteoclast generation or induce RANKL expression by osteoblasts.[27] IL-1 also stimulates TRAF6 expression, which leads to fusion of pre-osteoclasts and

differentiation.[21] Other cytokines—such as IL-4, IL-5, IL-10, interferon (IFN)-α, IFN-β, and IFN-γ—inhibit osteoclastogenesis by blocking RANKL signaling. IFN-γ also downregulates TRAF6, resulting in inhibition of osteoclast formation.[28]

Toll-like receptors (TLRs) are a class of receptors that recognize specific pathogen-associated molecular patterns, such as lipopolysaccharides (LPSs) on Gram-negative bacteria, and are involved in mediating the inflammatory response by activating transcription factors such as NF-κB and AP-1. TLRs also regulate osteoclast function. Activation of the TLR signaling pathway appears to not only inhibit early osteoclast precursors, but also stimulates mature osteoclasts as well.[29,30]

Osteopontin (OPN) is a multifunctional protein with cytokine properties produced by immune cells, osteoclasts, osteoblasts, endothelial cells, and epithelial cells. OPN levels increase in the presence of inflammatory disease, sepsis, and LPS administration. In the immune system, OPN has chemotactic properties, which recruit cells to inflammatory sites. OPN also regulates cytokine production and apoptosis. In bone, OPN mediates calcitriol-induced bone resorption and may suppress the anabolic activity of PTH on bone.[31–36]

In addition to promoting osteoclastogenesis, cytokines also inhibit new bone formation by inhibiting osteoblast development. The wingless-type mouse mammas tumor virus (MMTV) integration site (Wnt)/β-catenin pathway increases bone formation by regulating osteoblast differentiation, proliferation, and apoptosis. Wnt binds to the cell surface receptor complex low-density lipoprotein (LDL) related proteins (Lrp5 and Lrp6) and their coreceptor frizzled to activate the canonical signaling pathway.[37] Activation of Wnt signaling is important for the regulation of the inflammatory response and may be a key element in the link between inflammation and bone metabolism. Through activation of the intracellular protein disheveled and use of axin, a scaffolding protein, the enzyme glycogen synthase-3β (GSK-3β) is phosphorylated and inhibited. Under normal circumstances, GSK-3β phosphorylates β-catenin, thereby targeting it for ubiquination and degradation and hence regulating intracellular β-catenin levels. Unphosphorylated β-catenin is free to translocate to the nucleus and regulate the transcription of genes involved in the differentiation of mesenchymal cells into osteoblasts, rather than adipocytes. Activation of this pathway through Lrp5 is reported to be important in responding to changes in mechanical load.[38] TNF-α inhibits Wnt signaling through inducton of dickkopf 1. OPN may also inhibit Wnt signaling, possibly through PTH suppression.[39] The Wnt signal pathway is increasingly recognized as an important link between inflammation and bone remodeling through its interaction with osteoblasts, cytokines, and PTH.[40,41]

In humans, the relationship between inflammation and bone loss has mainly been demonstrated in the setting of chronic inflammatory rheumatologic diseases, rather than in critical illness. However, one study of critically ill individuals revealed that these patients have increased biochemical markers of bone resorption in the setting of persistently elevated cytokine levels, including IL-6, IL-1, or TNF-α.[22]

To examine the effects of acute infection and inflammation on the markers of bone turnover, the bacterial endotoxin, LPS, was administered to 10 healthy human volunteers in a placebo-controlled cross-over trial.[33] Administration of LPS induced a transient decrease in PTH levels and an increase in OPN levels. The biochemical markers of bone metabolism revealed a decrease in CTX and an increase in P1NP, with no change in OCN. These results were suggestive of increased activity of immature osteoblasts and perhaps impaired maturation of osteoblasts. Acutely, LPS appeared to shift the balance toward bone formation, with a decrease in the lytic activity of osteoclasts. The initial increase in bone formation resonates with activation of the INA observed in ACI before the subsequent downregulation seen in PACI–CCI.[33]

## Nutrition and bone disease

Apart from immobilization and inflammation, nutritional deficiencies are being increasingly recognized as a major contributor to metabolic bone disease in CCI.

### Vitamin D deficiency

Vitamin D insufficiency and frank deficiency are highly prevalent (91–97%) in individuals with CCI.[14,22,42] In one study of 22 CCI patients, mean 25-hydroxy vitamin D levels were 11 ng/mL.[22] In addition, these individuals had low total and free 1,25 vitamin D levels, potentially due to renal disease, cytokine release, and GH/IGF-1 deficiency, impairing the activity of 1α-hydroxylase.[43–45] The CCI patients in the above cohort had pronounced bone resorption as part of the catabolic state. Although the bone formation markers, PICP and PINP, were also increased, it was suggested that an osteoblast maturation defect leads to low activity of mature osteoblasts, as indicated by sALP and OCN concentrations. In this study, secondary hyperparathyroidism was not evident in the majority of patients.[22]

Low levels of vitamin D–binding protein (DBP) were also found and correlated with ICU mortality.[46] Actin, which is released by organ trauma or dysfunction, has been shown to bind DBP and accelerate DBP and vitamin D clearance in animal models.[47] It has been argued that vitamin D levels in CCI patients may be misleading due to suppression of DBP, hypoalbuminuria, and vitamin D binding and production by immune cells; therefore, elevations in PTH may be a better indicator of vitamin D deficiency than vitamin D levels in this population.[14]

The high prevalence of vitamin D deficiency in CCI patients also reflects premorbid vitamin D deficiency in this population. In addition, lack of sunlight, inadequate dietary intake, malabsorption, hepatic dysfunction, renal dysfunction, and increased clearance of vitamin D may account for the high prevalence of vitamin D insufficiency/deficiency.[14]

### Other nutritional deficiencies and bone loss

Although vitamin D is the best studied nutrient affecting bone remodeling in CCI patients, other nutritional factors may also contribute to bone loss. The electrolytes magnesium and potassium stabilize the alkaline environment of bone and are protective from degradation.[48] Magnesium is also incorporated into bone mineral and is believed to contribute to bone strength.[49] Animal studies have revealed that magnesium deficiency is associated with decreased OPG and increased RANKL, contributing to increased osteoclastogenesis.[50] Studies in non-CCI patients have demonstrated a protective effect of magnesium and potassium on bone.[51,52] Potassium intake and increased urinary potassium excretion is associated with decreased urinary calcium loss and increased bone mineral density in elderly postmenopausal women.[53]

Vitamin C deficiency has also been associated with inhibition of bone resorption. Vitamin C is an essential cofactor for collagen formation and may reduce oxidative stress and inhibit bone resorption. However, observational studies of non-CCI patients, such as the Framingham Osteoporosis Study, the Women's Health Initiative, and data from National Health and Nutrition Examination Survey (NHANES) III, have had variable results on the effects of vitamin C intake and ascorbic acid levels on fracture incidence.[54–56] The relation of vitamin C deficiency and fracture in these population studies may be confounded by the obesity epidemic. Obese individuals demonstrate high rates of vitamin C deficiency,[57] yet their bone mass is increased and they demonstrate lower fracture incidence, which is thought to result from increased weight bearing in daily activities.[58]

Vitamin K is also important for carboxylation of proteins in bone. In the Framingham cohort and more recently in a prospective study of Japanese women, increased dietary vitamin K intake has been shown to reduce hip fractures, whereas deficiency increased the risk of vertebral fractures.[59–61]

Dietary carotenoids may protect the bone from resorption by reducing oxidative stress. The Framingham Osteoporosis Study showed positive correlations between carotenoids and bone mineral density. Total carotenoids, β-carotene, lycopene, and lutein plus zeaxanthin have been inversely associated with 4-year loss in bone mineral density.[62] *In vitro* studies have suggested that lycopene may inhibit osteoclast formation, while stimulating osteoblasts.[63] Recent observational studies, including the Framingham Offspring Study and NHANES III, have shown a correlation between levels of vitamin B12, bone mineral density, and fracture risk.[64,65] This observed association may be mediated through the high homocysteine levels seen in vitamin B12 deficiency, which result in impaired collagen cross-linking and may increase fracture risk.[48] Folate deficiency was associated with hip fracture risk in the Hordaland Homocysteine Study and with low bone mineral density in the Framingham Osteoporosis Study.[66] Vitamin B6 deficiency has also been correlated with risk of fracture, in the Rotterdam Study.[67] Supplementing folic acid and vitamin B12 in a Japanese population with high homocysteine levels was seen to reduce hip fractures by 80%.[68]

Macronutrient intake remains an important consideration in CCI patients. Both adequate calorie and protein intake are associated with improved bone mineral density.[69–72] The classic hypothesis that high dietary protein may be detrimental to bone mineralization and calcium homeostasis seems to be inconsistent with these epidemiologic studies as well as with several trials in healthy postmenopausal women that demonstrate increased calcium retention with high protein intake.[73,74]

These nutritional factors play an important role in bone metabolism in CCI, as these patients have high rates of micronutrient deficiencies and will require enteral or parenteral nutritional supplementation.[75]

## Hypothalamic–pituitary function

Allostatic set-point changes in the INA in ACI represent an adaptive response in order for the individual to survive. Failure to downregulate this INA stress response will lead to further muscle breakdown, immunosuppression, and bone resorption, known as allostatic overload.[76] The most important hormonal responses with regard to bone disease in this state of allostatic overload are the ACTH/cortisol, GH/IGF-1, gonadal, and thyroid axes. Initially, cortisol, GH, TSH, and cytokines are elevated in ACI, but as the individual enters PACI, hypothalamic–pituitary axis activity is suppressed.[77,78] In PACI/CCI, plasma cortisol levels remain elevated due to direct stimulation of the adrenal gland by humoral factors other than ACTH.[79]

ACI is associated with a rise in cortisol levels due to increased ACTH and CRH release. This may occur due to direct stimulation of CRH and ACTH release or by resistance to the negative feedback by cortisol.[4] The diurnal variation of cortisol is lost, and adrenal cortisol production is directly stimulated by cytokines including IL-6 and endothelin-1.[80–82] In addition, cortisol-binding globulin (CBG) levels are decreased in inflammatory states via cleavage by neutrophil elastase, resulting in higher levels of free cortisol.[4] This hypercortisolism in ACI is beneficial, as it alters carbohydrate, fat, and protein metabolism; suppresses excessive inflammation; and promotes renal salt and water retention to improve hemodynamic status.[4] In CCI, cortisol levels remain elevated, but CBG levels increase and ACTH levels decline. This functional hypercortisolism likely contributes to metabolic bone disease, but whether this chronic increase in cortisol is beneficial or detrimental to the individual as a whole in CCI is unknown.[4,79]

GH levels are elevated in ACI; however, IGF-1 levels are low due to impaired GH signal transduction, resulting in decreased transcription of IGF-1 and GH-dependent IGF-binding proteins.[83] This pattern results in the insulin-antagonizing effects of GH (increased serum glucose and free fatty acid levels) and loss of the anabolic effects of IGF-1 signaling.[4] In PACI, GH pulsatility decreases, although the nonpulsatile levels may remain elevated, and IGF-1 levels remain low, through mechanisms believed to be mediated by the hypothalamus.[7,84] GH receptors are present on chondrocytes and osteoblasts and mediate the growth of osteoblasts by phosphorylation of the Janus tyrosine kinase/signal transducers and activator of transcription pathway and the MAPK pathway.[85,86] In addition, GH increases 1,25-D levels.[87] Murine models studying the function of IGF-1 and bone have demonstrated the importance of IGF-1 for bone mineralization, osteoblast differentiation, and regulation of OPG and RANK ligand.[86,88,89] Therefore, the low levels of IGF-I in CCI may increase bone loss.

Hypothalamic–pituitary–thyroid axis activity changes from ACI to PACI–CCI. During the initial phase of ACI, there is a rapid fall in T3 levels, with a rise in reverse T3 (rT3), due to decreased type 1 deiodinase and increased type 3 deiodinase, leading to increased conversion of T4 to rT3.[90] Changes in TSH secretion also occur, with decreased TSH pulsatility and loss of the nocturnal TSH surge.[6] The suggested mechanism has been related to cytokine release; low thyroid hormone-binding proteins; and inhibition of thyroid hormone binding, transport, and metabolism due to elevated levels of free fatty acids and bilirubin.[4] With PACI, TSH secretion is significantly reduced. Animal models of prolonged critical illness have demonstrated decreased *TRH* gene expression in the hypothalamus.[91] Thyroid hormone levels in critical illness are inversely correlated with markers of bone degradation.[90] TSH itself may also have an important suppressive role in osteoclast function.[92] Activation of TSH receptors on the surface of osteoclasts reduces resorption of bone. The low levels of TSH that are characteristic of CCI have the potential to increase osteoclast-mediated bone resorption.

The gonadal axis is also affected by critical illness, with an immediate fall in testosterone levels in men during ACI, despite elevated luteinizing hormone (LH) levels. This may be in part due to decreased testosterone production, in addition to increased testosterone clearance by peripheral aromatization of testosterone to estrogen.[93] For both critically ill men and women, estradiol levels are low; however, a relative increase in estradiol is associated with decreased survival in critical illness.[94] During PACI, testosterone levels remain low and LH concentrations also fall, along with sex hormone-binding globulin. The hypogonadism of critical illness is only partly responsive to exogenous gonadotropin-releasing hormone, thus there appear to be both central and peripheral defects in the male gonadal axis.[95] Hypogonadism is also known to adversely affect bone density. Therefore, hypercortisolism, decreased IGF-I, hypothyroidism, including low TSH, and hypogonadism in CCI may all contribute to abnormal bone remodeling in CCI.[96]

## Therapeutics

Given the high frequency of vitamin D deficiency in CCI patients and the consequences of vitamin D deficiency—not only on bone, but also on muscle wasting and weakness—routine supplementation is recommended in those without frank hypercalcemia or hypercalciuria.[3] The optimal dose in these patients is not known, but doses of up to 600 units daily have been shown to be inadequate.[22] Therefore, ergocalciferol 2000 IU daily should be considered to achieve a target 25-hydroxyvitamin D level of >32 ng/mL. As discussed, CCI patients may have low 1α-hydroxylase activity. Therefore, in those without hypercalcemia or hypercalciuria, additional supplementation with activated vitamin D (calcitriol 0.25 μg daily) should also be considered. In addition, calcium supplementation should be instituted to compensate for the urinary calcium losses. As discussed, elevated homocysteine levels, associated with certain vitamin B-complex deficiencies, may also be linked to metabolic bone disease and increased fracture risk. Adequate folic acid, vitamin B12, possibly vitamin B6 supplementation, and adequate protein-calorie administration are recommended.[97] Early mobilization through physical therapy may prove beneficial to bone metabolism and should be considered in critically ill patients.[98]

Bisphosphonate therapy inhibits osteoclast recruitment and function and so inhibits bone resorption, in addition to decreasing the circulating levels of osteoclast stimulators such as IL-6.[99] Administering intravenous (i.v.) pamidronate, a potent bisphosphonate, to CCI patients decreases markers of bone turnover at an average follow-up of 26 days.[100] Use of i.v. ibandronate, which has lower affinity for farnesyl diphosphate synthase, only transiently decreases CTX in CCI patients over a 6-day period (unpublished data). CCI patients, particularly those with end-stage renal disease on hemodialysis, should be evaluated for adynamic bone disease, prior to administration of i.v. bisphosphonates. Elevated NTX, appropriately elevated PTH, and/or nonsuppressed OCN levels make adynamic bone disease unlikely and, therefore, would favor the administration of a bisphosphonate to decrease bone hyperresorption. Other osteoporosis therapies given in healthy individuals such as calcitonin, teriparatide, and denosumab are also potentially beneficial in CCI patients, although their use has not been studied in this population.

## Conclusions

CCI is increasing in prevalence due to the ability of medical care, especially technology, to support high-risk critically ill patients who have survived ACI and PACI. In these prolonged critical illness stages, physiological mechanisms are activated that have not had the benefit of natural selection and therefore may not be entirely beneficial.

The severe bone loss observed in CCI is not due to immobilization, altered hormone levels, nutritional deficiencies, or inflammation alone, but likely due to a combination of these factors. Apart from the risk of bone fractures associated with bone hyperresorption, bone resorption can cause hypercalciuria, water loss and dehydration, nephrolithiasis, and nephrocalcinosis, and potentially increased mortality. Given the high mortality rates in the six months following discharge from hospital after a CCI, treating the metabolic and endocrine abnormalities that lead to hyperresorption of bone may not only reduce future fracture risk, but also may improve the survival for these patients. As the annual number of cases of individuals who develop CCI is expected to more than double by 2020, studies should address how to improve the long-term morbidity and mortality of these patients. Understanding the allostatic mechanisms involved in critical illness and discovering whether these mechanisms contribute to the pathophysiology of CCI may improve long-term survival of this fragile population.

## Conflicts of interest

The authors declare no conflicts of interest.

## References

1. Zorab, J. 2003. Bjørn Ibsen. *Resuscitation* **57**: 3–9.
2. Wiencek, C. & C. Winkelman. 2010. Chronic critical illness: prevalence, profile, and pathophysiology. *AACN Adv. Crit. Care* **21**: 44–61. quiz 63.
3. Hollander, J.M. & J.I. Mechanick. 2006. Nutrition support and the chronic critical illness syndrome. *Nutr. Clin. Pract.* **21**: 587–604.
4. Vanhorebeek, I., L. Langouche & G. Van den Berghe. 2006. Endocrine aspects of acute and prolonged critical illness. *Nat. Clin. Pract. Endocrinol. Metab.* **2**: 20–31.
5. Via, M., C. Scurlock, D. Adams, *et al.* 2010. An impaired postoperative hyperglycemic stress response is associated with increased mortality in cardiothoracic surgery patients. *Endocr. Pract.* **29**: 1–17.
6. Van den Berghe, G. *et al.* 1998. Neuroendocrinology of prolonged critical illness: effects of exogenous thyrotropin-releasing hormone and its combination with growth hormone secretagogues. *J. Clin. Endocrinol. Metab.* **83**: 309–319.
7. Van den Berghe, G., F. de Zegher & R. Bouillon. 1998. Clinical review 95: acute and prolonged critical illness as different neuroendocrine paradigms. *J. Clin. Endocrinol. Metab.* **83**: 1827–1834.
8. Weekers, F. *et al.* 2002. A novel *in vivo* rabbit model of hypercatabolic critical illness reveals a biphasic neuroendocrine stress response. *Endocrinology* **143**: 764–774.
9. Girard, K. & T.A. Raffin. 1985. The chronically critically ill: to save or let die? *Respir. Care* **30**: 339–347.
10. Carson, S.S. & P.B. Bach. 2002. The epidemiology and costs of chronic critical illness. *Crit. Care Clin.* **18**: 461–476.
11. Douglas, S. *et al.* 1996. Survival experience of chronically critically ill patients. *Nurs. Res.* **45**: 73–77.
12. Nasraway, S.A. Jr., T.M. Hudson-Jinks & R.M. Kelleher. 2002. Multidisciplinary care of the obese patient with chronic critical illness after surgery. *Crit. Care Clin.* **18**: 643–657.
13. Nierman, D.M. 2002. A structure of care for the chronically critically ill. *Crit. Care Clin.* **18**: 477–491.
14. Nierman, D.M. & J.I. Mechanick. 1998. Bone hyperresorption is prevalent in chronically critically ill patients. *Chest* **114**: 1122–1128.
15. Nelson, J.E. *et al.* 2004. The symptom burden of chronic critical illness. *Crit. Care Med.* **32**: 1527–1534.

16. Zilberberg, M.D., M. de Wit, J.R. Pirone & A.F. Shorr. 2008. Growth in adult prolonged acute mechanical ventilation: implications for healthcare delivery. *Crit. Care Med.* **36:** 1451–1455.

17. Hollander, J.M. & J.I. Mechanick. 2009. Bisphosphonates and metabolic bone disease in the ICU. *Curr. Opin. Clin. Nutr. Metab. Care* **12:** 190–195.

18. Mechanick, J.I. *et al.* 1997. Parathyroid hormone suppression in spinal cord injury patients is associated with the degree of neurologic impairment and not the level of injury. *Arch. Phys. Med. Rehabil.* **78:** 692–696.

19. Sambrook, P.N. *et al.* 2006. High bone turnover is an independent predictor of mortality in the frail elderly. *J. Bone Miner. Res.* **21:** 549–555.

20. Hadjidakis, D.J. & I.I. Bone Androulakis. 2006. Remodeling. *Ann. N.Y. Acad. Sci.* **1092:** 385–396.

21. Blair, H. & M. Zaidi. 2006. Osteoclastic differentiation and function regulated by old and new pathways. *Rev. Endocr. Metab. Disord.* **7:** 23–32.

22. Van den Berghe, G. *et al.* 2003. Bone turnover in prolonged critical illness: effect of vitamin D. *J. Clin. Endocrinol. Metab.* **88:** 4623–4632.

23. Caetano-Lopes, J., H. Canhão & J.E. Fonseca. 2009. Osteoimmunology—the hidden immune regulation of bone. *Autoimmun. Rev.* **8:** 250–255.

24. Asagiri, M. & H. Takayanagi. 2007. The molecular understanding of osteoclast differentiation. *Bone* **40:** 251–264.

25. Takayanagi, H. 2007. Osteoimmunology: shared mechanisms and crosstalk between the immune and bone systems. *Nat. Rev. Immunol.* **7:** 292–304.

26. Pasquale, E.B. 2008. Eph-ephrin bidirectional signaling in physiology and disease. *Cell* **133:** 38–52.

27. Nanes, M.S. 2003. Tumor necrosis factor-alpha: molecular and cellular mechanisms in skeletal pathology. *Gene* **321:** 1–15.

28. Datta, H.K., W.F. Ng, J.A. Walker, *et al.* 2008. The cell biology of bone metabolism. *J. Clin. Pathol.* **61:** 577–587.

29. Bar-Shavit, Z. 2008. Taking a toll on the bones: regulation of bone metabolism by innate immune regulators. *Autoimmunity* **41:** 195–203.

30. Hurst, J. & P. von Landenberg. 2008. Toll-like receptors and autoimmunity. *Autoimmun. Rev.* **7:** 204–208.

31. Choi, S.T. *et al.* 2008. Osteopontin might be involved in bone remodelling rather than in inflammation in ankylosing spondylitis. *Rheumatology (Oxford)* **47:** 1775–1779.

32. Denhardt, D.T., M. Noda, A.W. O'Regan, *et al.* 2001. Osteopontin as a means to cope with environmental insults: regulation of inflammation, tissue remodeling, and cell survival. *J. Clin. Invest.* **107:** 1055–1061.

33. Grimm, G. *et al.* 2010. Changes in osteopontin and in biomarkers of bone turnover during human endotoxemia. *Bone.* **47:** 388–391.

34. Ono, N. *et al.* 2008. Osteopontin negatively regulates parathyroid hormone receptor signaling in osteoblasts. *J. Biol. Chem.* **283:** 19400–19409.

35. Wang, K.X. & D.T. Denhardt. 2008. Osteopontin: role in immune regulation and stress responses. *Cytokine Growth Factor Rev.* **19:** 333–345.

36. Reinholt, F.P., K. Hultenby, A. Oldberg & D. Heinegard. 1990. Osteopontin–a possible anchor of osteoclasts to bone. *Proc. Natl. Acad. Sci. USA* **87:** 4473–4475.

37. Zaidi, M. 2007. Skeletal remodeling in health and disease. *Nat. Med.* **13:** 791–801.

38. Clevers, H. 2006. Wnt/[beta]-catenin signaling in development and disease. *Cell* **127:** 469–480.

39. Wai, P. & P. Kuo. 2008. Osteopontin: regulation in tumor metastasis. *Cancer Metastasis Rev.* **27:** 103–118.

40. Johnson, M.L. & M.A. Kamel. 2007. The Wnt signaling pathway and bone metabolism. *Curr. Opin. Rheumatol.* **19:** 376–382. 310.1097/BOR.1090b1013e32816e32806f32819.

41. Krishnan, V., H.U. Bryant & O.A. MacDougald. 2006. Regulation of bone mass by Wnt signaling. *J Clin Invest* **116:** 1202–1209.

42. Lee, P., J.A. Eisman & J.R. Center. 2009. Vitamin D deficiency in critically ill patients. *N. Engl. J. Med.* **360:** 1912–1914.

43. Van den Berghe, G. 2001. The neuroendocrine response to stress is a dynamic process. *Best. Pract. Res. Clin. Endocrinol. Metab.* **15:** 405–419.

44. Van den Berghe, G. 2002. Neuroendocrine pathobiology of chronic critical illness. *Crit. Care Clin.* **18:** 509–528.

45. Van den Berghe, G. 2003. Endocrine evaluation of patients with critical illness. *Endocrinol. Metab. Clin. North Am.* **32:** 385–410.

46. Dahl, B. *et al.* 1999. Admission level of Gc-globulin predicts outcome after multiple trauma. *Injury* **30:** 275–281.

47. Goldschmidt-Clermont, P.J. *et al.* 1988. Role of group-specific component (vitamin D binding protein) in clearance of actin from the circulation in the rabbit. *J. Clin. Invest.* **81:** 1519–1527.

48. Tucker, K. 2009. Osteoporosis prevention and nutrition. *Curr. Osteoporos. Rep.* **7:** 111–117.

49. Rude, R., F. Singer & H. Gruber. 2009. Skeletal and hormonal effects of magnesium deficiency. *J. Am. Coll. Nutr.* **28:** 131–141.

50. Rude, R., H. Gruber, L. Wei & A. Frausto. 2005. Immunolocalization of RANKL is increased and OPG decreased during dietary magnesium deficiency in the rat. *Nutr. Metab* **2:** 24.

51. New, S.A., S.P. Robins & M.K. Campbell. 2000. Dietary influences on bone mass and bone metabolism: further evidence of a positive link between fruit and vegetable consumption and bone health? *Am. J. Clin. Nutr.* **71:** 142–151.

52. Tucker, K.L., M.T. Hannan & H. Chen. 1999. Potassium, magnesium and fruit and vegetable intakes are associated with greater bone mineral density in elderly men and women. *Am. J. Clin. Nutr.* **69:** 727–736.

53. Zhu, K., A. Devine & R. Prince. 2009. The effects of high potassium consumption on bone mineral density in a prospective cohort study of elderly postmenopausal women. *Osteoporos. Int.* **20:** 335–340.

54. Sahni, S. *et al.* 2008. High vitamin C intake is associated with lower 4-year bone loss in elderly men. *J. Nutr.* **138:** 1931–1938.

55. Simon, J.A. & E.S. Hudes. 2001. Relation of ascorbic acid to bone mineral density and self-reported fractures among US adults. *Am. J. Epidemiol.* **154:** 427–433.

56. Wolf, R.L. *et al.* 2005. Lack of a relation between vitamin and mineral antioxidants and bone mineral density: results from the women's health initiative. *Am. J. Clin. Nutr.* **82:** 581–588.

57. Kaidar-Person, O., B. Person, S. Szomstein & R.J. Rosenthal. 2008. Nutritional deficiencies in morbidly obese patients: a new form of malnutrition?: part A: vitamins. *Obes. Surg.* doi:10.1007/511695-007-9349-4.

58. Pesonen, J. *et al.* 2005. High bone mineral density among perimenopausal women. *Osteoporos. Int.* **16:** 1899–1906.

59. Booth, S.L. *et al.* 2000. Dietary vitamin K intakes are associated with hip fracture but not with bone mineral density in elderly men and women. *Am. J. Clin. Nutr.* **71:** 1201–1208.

60. Feskanich, D. *et al.* 1999. Vitamin K intake and hip fractures in women: a prospective study. *Am. J. Clin. Nutr.* **69:** 74–79.

61. Tsugawa, N. *et al.* 2008. Low plasma phylloquinone concentration is associated with high incidence of vertebral fracture in Japanese women. *J. Bone Miner. Metab.* **26:** 79–85.

62. Sahni, S. *et al.* 2009. Protective effect of total and supplemental vitamin C intake on the risk of hip fracture—a 17-year follow-up from the Framingham osteoporosis study. *Osteoporos. Int.* **20:** 1853–1861.

63. Kim, L., A.V. Rao & L.G. Rao. 2003. Lycopene II–effect on osteoblasts: the carotenoid lycopene stimulates cell proliferation and alkaline phosphatase activity of SaOS-2 cells. *J. Med. Food* **6:** 79–86.

64. Morris, M.S., P.F. Jacques & J. Selhub. 2005. Relation between homocysteine and B-vitamin status indicators and bone mineral density in older Americans. *Bone* **37:** 234–242.

65. Tucker, K.L. *et al.* 2005. Low plasma vitamin B12 is associated with lower BMD: the Framingham osteoporosis study. *J. Bone Miner. Res.* **20:** 152–158.

66. McLean, R.R. *et al.* 2004. Homocysteine as a predictive factor for hip fracture in older persons. *N. Engl. J. Med.* **350:** 2042–2049.

67. Yazdanpanah, N. *et al.* 2007. Effect of dietary B vitamins on BMD and risk of fracture in elderly men and women: the rotterdam study. *Bone* **41:** 987–994.

68. Sato, Y., Y. Honda, J. Iwamoto, *et al.* 2005. Effect of folate and mecobalamin on hip fractures in patients with stroke: a randomized controlled trial. *JAMA* **293:** 1082–1088.

69. Promislow, J.H., D. Goodman-Gruen, D.J. Slymen & E. Barrett-Connor. 2002. Protein consumption and bone mineral density in the elderly: the Rancho Bernardo study. *Am. J. Epidemiol.* **155:** 636–644.

70. Kerstetter, J.E., A.C. Looker & K.L. Insogna. 2000. Low dietary protein and low bone density. *Calcif. Tissue Int.* **66:** 313.

71. Rapuri, P.B., J.C. Gallagher & V. Haynatzka. 2003. Protein intake: effects on bone mineral density and the rate of bone loss in elderly women. *Am. J. Clin. Nutr.* **77:** 1517–1525.

72. Hannan, M.T. *et al.* 2000. Effect of dietary protein on bone loss in elderly men and women: the Framingham osteoporosis study. *J. Bone Miner. Res.* **15:** 2504–2512.

73. Hunt, J.R., L.K. Johnson & Z.K. Fariba Roughead. 2009. Dietary protein and calcium interact to influence calcium retention: a controlled feeding study. *Am. J. Clin. Nutr.* **89:** 1357–1365.

74. Kerstetter, J.E., K.O. O'Brien, D.M. Caseria, *et al.* 2005. The impact of dietary protein on calcium absorption and kinetic measures of bone turnover in women. *J. Clin. Endocrinol. Metab.* **90:** 26–31.

75. Corcoran, T.B., M.A. O'Neill, S.A. Webb & K.M. Ho. 2009. Prevalence of vitamin deficiencies on admission: relationship to hospital mortality in critically ill patients. *Anaesth. Intensive Care* **37:** 254–260.

76. McEwen, B.S. 1998. Stress, adaptation, and disease. Allostasis and allostatic load. *Ann. N.Y. Acad. Sci.* **840:** 33–44.

77. Bornstein, S.R. & G.P. Chrousos. 1999. Clinical review 104: adrenocorticotropin (ACTH)- and non-ACTH-mediated regulation of the adrenal cortex: neural and immune inputs. *J. Clin. Endocrinol. Metab.* **84:** 1729–1736.

78. Van den Berghe, G.H. 1999. The neuroendocrine stress response and modern intensive care: the concept revisited. *Burns* **25:** 7–16.

79. Vermes, I. & A. Beishuizen. 2001. The hypothalamic–pituitary–adrenal response to critical illness. *Best Pract. Res. Clin. Endocrinol. Metab.* **15:** 495–511.

80. Beishuizen, A. & L.G. Thijs. 2001. Relative adrenal failure in intensive care: an identifiable problem requiring treatment? *Best Pract. Res. Clin. Endocrinol. Metab.* **15:** 513–531.

81. Bornstein, S.R., W.C. Engeland, M. Ehrhart-Bornstein & J.P. Herman. 2008. Dissociation of ACTH and glucocorticoids. *Trends Endocrinol. Metab.* **19:** 175–180.

82. Cooper, M. & P. Stewart. 2003. Corticosteroid insufficiency in acutely ill patients. *N. Engl. J. Med.* **348:** 727–734.

83. Baxter, R.C. 2001. Changes in the IGF-IGFBP axis in critical illness. *Best Pract. Res. Clin. Endocrinol. Metab.* **15:** 421–434.

84. Van den Berghe, G. *et al.* 1999. Growth hormone-releasing peptide-2 infusion synchronizes growth hormone, thyrotrophin and prolactin release in prolonged critical illness. *Eur. J. Endocrinol.* **140:** 17–22.

85. Argetsinger, L.S. *et al.* 1993. Identification of JAK2 as a growth hormone receptor-associated tyrosine kinase. *Cell* **74:** 237–244.

86. Giustina, A., G. Mazziotti & E. Canalis. 2008. Growth hormone, insulin-like growth factors, and the skeleton. *Endocr. Rev.* **29:** 535–559.

87. Wei, S. *et al.* 1997. Growth hormone increases serum 1,25-dihydroxyvitamin D levels and decreases 24,25-dihydroxyvitamin D levels in children with growth hormone deficiency. *Eur. J. Endocrinol.* **136:** 45–51.

88. Clemens, T.L. & S.D. Chernausek. 2004. Genetic strategies for elucidating insulin-like growth factor action in bone. *Growth Horm. IGF Res.* **14:** 195–199.

89. Rubin, J. *et al.* 2002. IGF-I regulates osteoprotegerin (OPG) and receptor activator of nuclear factor-kappaB ligand in vitro and OPG in vivo. *J. Clin. Endocrinol. Metab.* **87:** 4273–4279.

90. Peeters, R.P., Y. Debaveye, E. Fliers & T.J. Visser. 2006. Changes within the thyroid axis during critical illness. *Crit. Care Clin.* **22:** 41–55. vi.

91. Mebis, L. *et al.* 2009. Expression of thyroid hormone transporters during critical illness. *Eur. J. Endocrinol.* **161:** 243–250.

92. Zaidi, M. *et al.* 2009. Thyroid-stimulating hormone, thyroid hormones, and bone loss. *Curr. Osteoporos Rep.* **2:** 47–52.

93. Spratt, D.I. *et al.* 2006. Increases in serum estrogen levels during major illness are caused by increased peripheral aromatization. *Am. J. Physiol. Endocrinol. Metab.* **291:** E631–E638.

94. May, A.K. *et al.* 2008. Estradiol is associated with mortality in critically ill trauma and surgical patients. *Crit. Care Med.* **36:** 62–68.

95. Van den Berghe, G. *et al.* 2001. Five-day pulsatile gonadotropin-releasing hormone administration unveils combined hypothalamic-pituitary-gonadal defects underlying profound hypoandrogenism in men with prolonged critical illness. *J. Clin. Endocrinol. Metab.* **86:** 3217–3226.

96. Nierman, D.M. & J.I. Mechanick. 1999. Hypotestosterone-mia in chronically critically ill men. *Crit. Care Med.* **27:** 2418–2421.

97. Demling, R. & M. DeBiasse. 1995. Micronutrients in critical illness. *Crit. Care Clin.* **11:** 651–673.

98. Needham, D.M. 2008. Mobilizing patients in the intensive care unit: improving neuromuscular weakness and physical function. *JAMA* **300:** 1685–1690.

99. Lissoni, P. *et al.* 1997. Acute effects of pamidronate administration on serum levels of interleukin-6 in advanced solid tumour patients with bone metastases and their possible implications in the immunotherapy of cancer with interleukin-2. *Eur. J. Cancer* **33:** 41–47.

100. Nierman, D.M. & J.I. Mechanick. 2000. Biochemical response to treatment of bone hyperresorption in chronically critically ill patients. *Chest* **118:** 761–766.

Ann. N.Y. Acad. Sci. ISSN 0077-8923

ANNALS OF THE NEW YORK ACADEMY OF SCIENCES

Issue: *Molecular and Integrative Physiology of the Musculoskeletal System*

# Integrative physiology of the aging bone: insights from animal and cellular models

Farhan A. Syed[1] and Kelley A. Hoey[2]

[1]Abbott Bioresearch Center, Worcester, Massachusetts. [2]Mayo Clinic, Rochester, Minnesota

Address for correspondence: Farhan A. Syed, Ph.D., Abbot Bioresearch Center, 100 Research Drive, Worcester, MA 01605-4314. farhan.syed@abbot.com

Age-related bone loss is a common worldwide phenomenon in the aging population, placing them at an increased risk of fractures. Fortunately, basic and translational studies have been pivotal in providing us with a mechanistic understanding of the cellular and molecular pathophysiology of this condition. This review focuses on the current concepts and paradigms of age-related bone loss and how various animal and cellular models have broadened our understanding in this fascinating but complex area. Changes in hormonal, neuronal, and biochemical cues with age and their effect on bone have been discussed. This review also outlines recent studies on the relationship between bone and fat in the marrow, as well as the fate of the marrow mesenchymal stromal cell population, which can give rise to either bone-forming osteoblasts or fat-forming adipocytic cells as a function of age.

Keywords: bone loss; aging; animal models

## Introduction

Human aging is a complex phenomenon wherein there is a progressive deterioration of function in a variety of organs and organ systems, including the skeletal system. The skeletal system is made up of specialized connective tissues, namely, the cartilage and bone. Both these tissues display distinct changes in structural and functional characteristics, as a function of age, leading to distinct pathologies such as arthritis involving primarily the cartilage and osteoporosis that is specific to bone.

Osteoporosis or "porous bone disease" is defined as a systemic skeletal disease characterized by low bone mass and micro-architectural deterioration of bone tissue, with a consequent increase in bone fragility and susceptibility to fracture. Osteoporosis is classified either as a primary or secondary disease based on whether other comorbid disorders or use of drugs which can cause osteoporosis are absent or present in a given patient.[1] Thus, traditionally "primary osteoporosis" refers to both postmenopausal osteoporosis in women and senile osteoporosis in both elderly men and women, which have been referred to as *Type I* and *Type II*, respectively.[1,2] Our understanding of the pathophysiology and gerontology of age-related bone loss has increased immensely from clinical observations and studies, as well as basic and translational research involving both cellular and animal models.

Bone is a dynamic organ with both mechanical and metabolic functions, and its composition reflects this role.[3] The bone composite comprises of organic and inorganic phases with its extracellular matrix consisting of hydroxyapatite mineral, collagenous and noncollagenous proteins, water, and also varying amounts of lipid, which depends on the age, species, and site.[3] At the cellular level, bone is composed of three structurally and functionally distinct cell types, namely, the bone-forming osteoblasts, the bone-resorbing osteoclasts, and the bone-regulating osteocytes. In the adult skeleton, osteoblasts account for 4–6%, osteoclasts for 1–2%, and osteocytes for more than 90% of the total bone cells.[4]

Bone is being constantly turned over in distinct areas of the skeleton, termed the bone metabolic units (BMUs) due to the activities of these cells.[5,6]

doi: 10.1111/j.1749-6632.2010.05813.x

In a healthy young adult skeleton, the processes of bone resorption and formation at a given BMU are tightly coupled, such that the rate of bone formation and matrix mineralization equals the rate of bone resorption and matrix degradation, thereby maintaining a steady state of bone.[6] However, during the process of aging, significant amounts of bone are lost due to the tipping of this delicate balance towards enhanced resorption coupled with decreased formation, leading to net bone loss in the aging population, which ultimately manifests as osteoporosis.[5] The reasons for this shift in balance are what inform our knowledge of age-related osteoporosis. Toward this end, a number of animal and cell models have been used in addition to human studies, which are beginning to shed light on the underlying cellular and molecular mechanisms.

We review in the following sections the various rodent and murine models that have been used in bone studies on aging and the information that has been forthcoming from them, especially from genetically engineered mice. We will also specifically outline recent studies that further our understanding of the cellular and molecular changes that occur in the bone and bone marrow with age, and how these are regulated either systemically or peripherally by various hormonal, biochemical, and neuronal cues.

## The aging bone: insights from large animal models

A number of animal models have been employed for studying age-related bone abnormalities. All these models have their specific advantages and limitations, which have to be considered when designing studies. Large and small animal models have been used to study age-related bone loss, including nonhuman primates, sheep, pigs, dogs, rats, and mice.

### Nonhuman primates

Nonhuman primates are probably the closest in their bone phenotype to humans, and hence offer an authentic model to study positive skeletal changes that occur with development, attainment of peak bone mass, as well as maintenance of the skeleton during adulthood and bone loss with age. For example, female cynomolgus monkeys display a reproductive and aging physiology closely mirroring humans.[7,8] Young skeletally immature (3–4 year old) cyanomolgous monkeys have high bone

turnover rates, whereas skeletally mature (10–15 year old) monkeys have decreasing bone biomarkers and bone formation rates, reflecting an overall decrease in bone turnover with age.[9] Additionally, the presence of cortical Haversian osteonal remodeling in nonhuman primates, as in humans, makes their use attractive as compared with rodents, which normally lack this phenomenon.[10] However, the prohibitive cost involved, significant handling and housing issues, and the long life span of these animals make their use rather limited, especially in a typical academic laboratory setting.

### Sheep, pigs, and dogs

In recent years, sheep have been increasingly employed as a model to study the efficacy of various agents, such as osseous implants and endogenous biochemicals, in preventing age-related or ovariectomy-induced osteoporosis.[11,12] For example, poor bone quality in ovariectomized or aged sheep compromised the osteointegrative efficacy of sandblasted titanium implants, both in trabecular and cortical bone components.[11] Aged sheep have also been used to test the efficacy of nonglycosylated BMP-2 coated implants in the tibia, and it was shown that such implants may foster bone healing and regeneration even in age-compromised individuals.[12]

Pigs have also been used as a model to test the efficacy of preclinical osseous and spinal implants to correct orthopedic abnormalities of aging.[13] Porcine mesenchymal stromal cells have been studied both *in vitro* and *ex-vivo* in order to determine the effects of age on osteogenic stromal cell commitment, which was found to decrease as a function of age.[14,15]

Dogs are often used in maxillofacial and orthopedic research to study decreases in bone formation and geriatric bone loss.[16,17] It has been recently demonstrated by bone labeling studies that old dogs display significant reductions in bone forming activity as compared with young dogs in the mandibular condyles.[16] Subtle differences were also observed in the bone regeneration brought about by cytokine TGF-beta 2 in older beagles.[17] Whereas aging did not seem to lessen the overall effectiveness of recombinant human TGF-beta 2 in enhancing regeneration of bone, thinner trabeculae and greater osteoid volume were observed as a function of age, pointing to decreases either in bone formation or lag in mineralization time in older animals.[17]

Such studies with larger animal models certainly underscore their importance in specific areas of geriatric research, especially as pre-clinical models for geriatric bone loss and therapy. Also, these models are substantially cheaper and friendlier to handle and groom compared with nonhuman primates. However, it needs to be pointed out that studies involving these animals can often be long drawn due to their relatively long life spans, and at times is cost prohibitive, especially if a larger cohort is required.

## Murine and rodent models of aging

### Normal aging models

An authentic model of senile osteoporosis can be argued to be normally aging mice or rats, which undergo bone loss as a function of age leading ultimately to the development of osteopenia and osteoporosis. It has been known for almost three decades that lipofuscin, an age pigment, accumulates in aging preosteoblasts, osteoblasts, and osteocytes in mice of either gender.[18] Therefore, in a number of studies, mice or rats of either gender aged normally without any interventions, in order to either cross-sectionally or longitudinally follow bone changes at the tissue, cellular, and molecular levels.[19,20] Commonly used laboratory mice have a life span of two to three years and attain peak bone mass between four and eight months of age; this is followed by a steady decline as the mice age.[21] For example, C57BL/6 mice exhibit decreased cancellous and cortical bone mass and quality as a function of age, displaying a senile osteoporotic phenotype.[19,20,22] There are subtle differences in the rate of cortical and cancellous bone loss in these aging mice, with the cancellous bone volume fraction decreasing from 6 weeks to 24 months while cortical thickness increases paralleling attainment of peak bone mass at around six months of age, which is then followed by a progressive decline.[19] This age-related loss in cortical thickness is also observed in both genders in humans, with age accounting for 52.9% of the variance in females and 37.9% in males.[23]

At the cellular level, there are changes in the osteoblastogenic and osteoclastogenic potential of their respective stem cells in the bone marrow. For example, 24-month-old BALB/c mice display decreases in proliferative potential of osteogenic stem cells.[24] A similar observation has recently been re-ported for aged C57BL/6 mice, which, in addition to decreases in the proliferation of their bone marrow stromal cells (BMSCs), also have significant impairment in their differentiation potential to osteoblasts.[25] Such studies seem to suggest that decreases in qualitative and quantitative potential of bone marrow stromal osteogenic stem cells can potentially lead to decreases in bone formation in these mice.[24,25] It has also been shown that hematopoiesis is affected by aging and mice are known to develop clonal B cell expansion and lymphoma with advanced age.[26–28] In this context, it is interesting to note that the gene expression of the osteoclast differentiation factor, receptor activator of nuclear factor kappaB ligand (RANKL)—which is important for osteoclastogenesis and is also secreted by B and T cells in addition to preosteoblasts—increases with age, and this correlates with changes in cancellous bone volume.[20] However, not all studies agree and actually show a decrease in the serum RANKL with aging in humans and rats.[29,30] Although it is generally recognized that aging is primarily associated with a decrease in osteoblasts, studies in aging mice have also reported increases in tartarate-resistant acid phosphatase (TRAP) positive osteoclasts with age.[30]

Male and female Wistar rats have been studied as models of normal aging, and they display progressive gain in bone parameters during growth and loss of bone density, both at trabecular and cortical sites along with cortical thinning after 12 months of age.[31,32] Interestingly, the magnitude of bone lost differed at different sites, at least in the females.[32]

Recently, Duque and colleagues have described the bone phenotype in a rat model of "healthy aging."[33] This LOU/c rat model displays an increased longevity in the absence of obesity and low incidence of age-related diseases. Using histomorphometric and biomarker assays for bone turnover, they were able to show in a cross-sectional study design that aging in these animals resulted in a reduction in the number of osteoblasts and osteoclasts along with a significant increase in the marrow adiposity of long bones. These changes at the tissue and cellular level correlated with decreases in serum C-telopeptide and serum osteocalcin, markers used for bone resorption and formation, respectively, whereas the levels of the calcitropic hormones, PTH and Vitamin D, remained unchanged with time.[33] This model thus offers promise in the future to study the mechanisms of senile osteoporosis in an otherwise healthy population.

It has to be noted that species and strain specific effects do remain. For example, decreases in bone formation seem to be a commonly observable phenomenon in both these experimental species; however, decreases in osteoclast number and function is observed in rats, but not in studies with mice.[20,24,25,30,33] Also, different inbred strains of mice display a range of peak bone phenotypes, which are genetically determined and are compartment specific. This can potentially impact the rate at which absolute amount of bone is lost with age.[34]

### Gonadectomized models

Gonadectomy is a generic term referring to the surgical removal of either the testes in males or the ovaries in females, which results in a loss of gonadal production of sex steroids. Sex steroids play a major role in the regulation of bone turnover. Thus, gonadectomy in either sex is associated with an increase in bone remodeling, increased bone resorption, and a relative deficit in bone formation, resulting in accelerated bone loss.[35] Therefore, these procedures have been routinely employed to induce models of bone loss and osteoporosis. It can be argued that these models cannot be classified as representing senile osteoporosis but rather mimic postmenopausal bone loss in women or castration in men. Nevertheless, they have been routinely used to understand mechanisms of bone loss with age, because sex steroid loss is also an important contributing factor to the overall loss of bone in the aging population.[2]

In mice and rodents, as in humans, the skeletal response to sex steroids is dose dependent. For example, low doses of estrogen prevent bone loss by reducing bone resorption, but higher doses can stimulate bone formation in mice.[36] It has also been suggested from human and animal studies involving estrogen receptor (ER) knockout mice that estrogens may be more protective than androgens in both men and women in preventing age-related bone loss.[37,38]

Estrogen or androgen loss can accelerate the effects of aging on bone by decreasing the antioxidant defense mechanisms in aging mice, thereby resulting in increase in oxidative stress and resultant damage to bone cells in C57BL/6 mice.[39] This has been achieved by comparing the phenotype and redox

status *per se,* in naturally aging bone versus bone obtained from gonadectomized mice. The results from this study indicate that loss of sex steroids can in fact accentuate age-related bone loss.[39]

Estrogen can also modulate the expression of other nuclear hormone receptors such as the vitamin D receptor (VDR). Ovariectomized young and old mice have been shown to have reductions in the expression of the VDR in osteoblasts, and estrogen supplementation in these mice actually reverses this phenotype more so in the old than in the young animals.[40] It is of note that ovariectomy in mice or rats leads to decreases in bone volume and increases in adipose volume in the bone marrow, a phenotype strikingly similar to that seen in aged humans.[41,42] It has also recently been established using paired iliac crest biopsies that marrow adipocyte volume can significantly increase in elderly osteoporotic postmenopausal women in a longitudinal fashion and estrogen clearly prevents this increase.[43]

We have recently shown, using an aging mouse model where relatively constant E levels were maintained in mice up to 22 months of age, that while trabecular bone loss was independent of E levels at least at the lumbar spine, cortical bone was maintained at levels comparable to six-month-old intact controls. This pattern of trabecular and cortical bone changes with age mirror what has been clinically observed in humans.[44] Such studies give us a strong rationale to consider the use of ovariectomized rodent and murine models as a surrogate for an aging bone phenotype, as long as the limitations are clearly understood and the complex interplay of sex hormone signaling is appreciated and taken into account when drawing valid conclusions from such models.

## Inbred and genetically modified mouse models

### SAMP6 mouse model of impaired bone formation

The classic model of senescence acceleration and age-associated disorders is the "senescence accelerated mouse," or the SAM mouse, originally developed by Takeda and colleagues in Japan.[45] A total of nine strains of accelerated aging were identified, out of which the SAMP6 strain was established as a model for senile osteoporosis characterized by low peak bone mass at their maturation.[46] The low bone mass in these mice is polygenetically determined, a situation akin to osteoporosis in humans.[36] It was initially shown that SAMP6 mice display both a reduction in cancellous bone formation rates due to a paucity of osteoblast progenitor cells in the marrow, and this phenotype becomes severe as these mice age.[47]

Recently, it has also been established that the impaired marrow osteogenesis in these mice is associated with reduction in cortical bone formation at the endocortical but not at the periosteal surface in long bones, and there seems to be a consistently low bone density across axial and appendicular bone sites as assessed by dual-energy X-ray absorptiometry.[48,49] An increase in bone resorption in these mice due to enhanced maturation of osteoclasts has also been reported.[50] However, it appears that this model is primarily one of the impaired bone formation as suggested by studies showing increased adipogenesis and myelopoiesis, and defective osteoblastogenesis and osteopenia.[51] That the defect is primarily in the osteoblastic cells and involves cytokines and intracellular pathways is also established by studies at the molecular level. For example, it has been shown that BMSCs derived from SAMP6 mice display reductions in the binding and activity of the AP-1 transcription factor Jun-D, and resultant decline in the expression of IL-11 which may impair bone formation.[52]

The Wnt pathway also seems to be suppressed in the osteoblasts of these mice because treatment with lithium chloride, which is a stimulator of the canonical Wnt pathway, enhances bone formation *in vivo*, and conversely the expression of secreted frizzled-related protein 4, an inhibitor of the Wnt pathway, is greatly enhanced in SAMP6 bone cells, thereby negatively regulating bone formation by inhibiting osteoblast proliferation and their responsiveness to ligands of the Wnt pathway such as Wnt 3A.[53,54]

## Klotho mouse: bone cells and mineral defects

The klotho gene-deficient mouse is also considered as an animal model for an accelerated aging state, mimicking a number of gerontological pathologies such as osteoporosis, skin atrophy, ectopic calcification, and gonadal dysplasia.[55] The klotho gene encodes a single pass transmembrane protein that forms a complex with multiple fibroblast growth factor (FGF) receptors and is a coreceptor for FGF23, which is a bone-derived hormone that induces a negative phosphate balance.[56] Klotho

deficiency causes irregular distributions of osteocytes and bone matrix proteins, as well as the accelerated aging of bone cells.[57]

Klotho gene-deficient mice display organelle-defective flat osteoblasts and small TRAP-positive osteoclasts at metaphyseal trabeculae in long bones.[58] A thick mineralized matrix in the proliferative zone of the growth plates and large non-mineralized areas in the metaphyseal trabeculae are also present, underscoring that both bone cellular and mineralization activities are severely impaired.[58] This study confirms an earlier report of impaired but independent osteoblastic and osteoclastic differentiation in these mice leading to a low turnover osteopenia.[59] Klotho functions as an aging suppressor and has been reported to be a regulator of oxidative stress and senescence, which increase with aging. In fact, loss of klotho leads to enhanced apoptosis in odontoblasts and disturbed mineralization in teeth.[60,61] In contrast to what had been described originally, it has been demonstrated by microcomputed tomography studies that in three-month-old klotho gene-deficient mice, the trabecular bone mass at the tibia actually increases over their wild-type litter mates.[54,62] Thus, there could be regional differences in the response to klotho's loss in the skeleton, which may well involve its ability to modulate pathways such as the Wnts and others involved in calcium and vitamin D homeostasis.[62] This model certainly warrants further study to further explore mechanisms of age-related osteoporosis.

## Laminopathies and extracellular matrix protein deficiencies

Mutations in the LMNA gene, which encodes for Lamin A, a component of the nuclear lamina, leads to at least a dozen distinct human diseases, collectively termed as "laminopathies," which primarily arise as a result of defects in the nuclear architecture.[51] Mice carrying an autosomal recessive mutation in the lamin A gene display defects and a premature aging phenotype similar to Werner's syndrome and Hutchinson–Gilford progeria syndrome in humans.[64] These mice display trabecular and cortical bone defects in the axial and appendicular skeleton consistent with an osteoporotic phenotype, which is also observed in human patients.[64] In order to elucidate the mechanisms of osteoporosis on loss of lamins, their expression was knocked down in normal human osteoblasts and in differentiating mesenchymal stem cells.[65] It was found that the loss of lamins was associated with lower osteoblast differentiation in BMSCs and a decrease in activity was also observed in mature osteoblasts. The decrease in osteoblastic commitment and differentiation was coupled to increased adipocyte differentiation in BMSCs which also displayed reduced Runx2 nuclear binding activity, although its expression was unchanged.[65] In agreement with these *in vitro* findings, mice knocked out for zmpste-24, a zinc metalloproteinase involved in the post-translational processing of prelamin A to mature lamin A, also displayed a progeroid phenotype in bone with a reduction in osteoblast and osteocyte numbers, and a parallel increase in marrow adipocytes.[66] Given the fact that oxidative stress increases with age and lamins are susceptible to free radical modifications (such as protein glycations) leading to genomic instability, it can be argued that loss of nuclear integrity and resultant modulation of transcription factor binding activity may play a role in the etiology of age-related bone loss due to reduced formation of osteoblasts.[67]

Extracellular matrix proteins play important roles in the regulation and signaling in bone cells.[68] Loss of short collagen IX of bone in male and female mice leads to significant loss of trabecular bone in an age-dependent fashion.[69] This progressive increase in resorption was accompanied by an increased RANKL/OPG ratio in osteoblasts, and structurally enlarged and active osteoclasts.[69] On the other hand, loss of osteopontin in mice leads to increases in bone formation *in vivo* and mineralized nodule formation *in vitro* on treatment with a prostaglandin E receptor agonist.[70] It is interesting to note in this regard that osteopontin expression in osteoblast decreases with aging at least in rats, suggesting the potential to design anabolic drugs if a similar decrease is found in aging humans.[71]

Biglycan is an extracellular matrix proteoglycan that is enriched in bone tissue. Biglycan-deficient mice develop age-dependent osteopenia and progressive osteoporosis, and this is related to decreases in the proliferation and commitment to osteoblasts coupled with an increased rate of apoptosis in BMSCs derived from these mice.[72,73] Changes in cellular matrix elasticity and integrity can be a potential mechanism of bone loss; for example, it has recently been shown *in vitro* that loss of RhoA signaling and cytoskeletal integrity in BMSCs can lead

to decreased osteogenesis and increased adipogenesis, a scenario similar to that observed in the aging marrow.[43,50,74]

## Peptides, hormones, and transcription factors

Genetically modified mice for various biochemicals such as hormones, peptides, and their cognate receptors and downstream transcription factors have yielded us with an appreciable amount of information on their roles in bone cell biology with implications for the aging skeleton. As already discussed, sex steroids play an important role in the aging skeleton.[37,38] For example, ER-alpha (ER$\alpha$) seems to protect the aging skeleton as indicated by the enhanced trabecular bone volume and increased expression of the osteogenic transcription factor cbfa-1/Runx2 in ER-beta knockout mice (ER$\beta^{-/-}$).[75]

The importance of the osteogenic transcription factor cbfa1/Runx2 is further highlighted by the inability of aged Runx2 haploinsufficient mice to respond to high-dose estrogen and also that aged mice require a full gene dosage of this transcription factor to recover cancellous bone volume after bone marrow ablation.[76,77] Transgenic mice overexpressing a dominant negative form of cbfa1/Runx2 ($\Delta$ cbfa1) postnatally, were also osteopenic due to reductions in bone formation rate brought about by lowered osteoblastic activity rather than numbers, suggesting that once osteoblasts are differentiated, cbfa1 can also regulate their function.[78] Conversely, mice haploinsufficient for the adipogenic transcription factor PPAR$\gamma$ display increased bone mass associated with increased osteoblastogenesis and decreased adipogenesis.[79]

Attenuation of bone morphogenetic protein pathways (BMPs) are also implicated in aging; for example, mice overexpressing the bone morphogenetic protein inhibitor noggin in mature osteoblasts showed dramatic decreases in bone mineral density and bone formation rates, suggesting that overproduction of noggin, a BMP pathway inhibitor during biological aging, may lead to the net bone loss.[80]

The role of growth hormone–IGF-1 axis has also been explored using knockout mice for these factors, and it is believed that deficiencies in this axis with age, due to reduced levels of both growth hormone and resultant decreases in IGF-1 levels, is another contributing factor for senile osteoporosis.[81]

## Central regulation of the aging skeleton

Bone remodeling is a homeostatic function such that the resorption of bone by osteoblasts is followed by formation by osteoblasts on bone surfaces. In recent years, there has been growing evidence that this homeostatic process of bone remodeling is under regulation via the central nervous system (CNS).[82] The first evidence that the CNS regulates bone remodeling was provided by experiments in the adipokine leptin-deficient *ob/ob* mice and mice lacking the signaling receptor to leptin (*db/db* mice).[83] It was established that leptin is a negative regulator of bone formation *in vivo* because the *ob/ob* and *db/db* mice displayed a high bone mass phenotype with increased rates of bone formation, although these mice were hypogonadal.[83] Further proof was provided by an elegant set of experiments where infusion of low doses of leptin intracerebroventricularly significantly decreased trabecular bone mass in *ob/ob* mice and wild type mice.[82]

Since the original description of a central regulation of bone remodeling, a number of studies have been carried out with genetically modified mice that have implicated a number of central neurotransmitters and receptors in the control of bone remodeling, such as cocaine and amphetamine regulated transcript, beta-adrenergic receptors, and cannabinoid receptors.[84–86] Recently, a direct role for cannabinoid receptors, which belong to the G-protein coupled receptor superfamily, has been shown in prevention of age-related osteoporosis using mice knocked out for the cannabinoid receptor (CB1$^{-/-}$ mice).[86] Young CB1 knockouts have a high bone mass phenotype, primarily due to reduced bone resorption; however, on aging, BMSCs from these mice display enhanced adipogenesis at the expense of osteogenesis by upregulating adipogenic signals such as cAMP and pCREB.[74,86] However, it needs to be mentioned that not all studies with the central regulation have yielded similar results. For example, a drop in serum leptin levels in aging mice is associated with decrease in bone mass and strength, and an increase in serum markers of bone resorption and bone phenotyping of axial and appendicular skeletal sites in leptin-deficient *ob/ob* mice established that loss of leptin leads to decreases in bone mineral density, cortical thickness, and trabecular volume in femurs with a parallel rise in adipocyte numbers. By contrast, at the lumbar vertebrae, increased trabecular bone parameters and decreased adipocyte numbers are observed, pointing to site specificity in leptin effects, and underscores the utility of this model in understanding age-related changes in marrow adipogenesis and osteopenia.[87,88] Also, a recent study using beta-adrenergic receptor knockout mice suggests direct positive effects of weight and/or leptin on bone turnover and cortical bone structure, independent of adrenergic signaling.[85]

## Changes in the aging bone marrow cells

Osteoclasts and osteoblasts are derived from two distinct cell lineages in the bone marrow. Osteoclasts are derived from hematopoietic stem cell precursors of the monocyte/macrophage lineage, whereas osteoblasts are derived from the BMSC compartment.[89,90] With aging, although an increase in bone resorption is seen, the primary pathophysiology of age-related bone loss as deduced from a number of histomorphometric evaluations of iliac crest biopsies obtained from aged individuals seems to be a drastic decrease in bone formation.[41,43,44] Thus, the age-related changes in osteoblast commitment, or relative lack of it from BMSCs, is of particular interest in terms of both the kind of molecular signals and the cellular pathways involved in order to understand the cellular and molecular basis of age-related osteoporosis.

Marrow stromal cells (MSCs) are a multipotent cell type that can potentially give rise not only to osteoblasts, but also to adipocytes, chondrocytes, and myoblasts.[91] The exact sequence of progression from a state of "stemness" to being a "progenitor," which is perhaps more advanced on the pathway to commitment, is not very clear.[92] However, a number of studies indicate that osteoblastic and adipocytic differentiation are closely linked, and it is believed that a bipotential osteoblast/adipocyte precursor exists whose commitment to either lineage depends on the strength of signals that favor either of these processes.[93] For example, the transcription factors Runx2 and osterix promote osteoblastic and C/EBPß, and PPARγ promote adipocytic commitment and differentiation.[78,94,95] That the balance between such lineage-specific nuclear transcription factors is important was highlighted by findings in mice with haplo-insufficiency of PPARγ; these mice have an increase in bone mass associated with increased osteoblastogenesis and decreased adipogenesis.[79]

Collectively, these findings have led to the concept of "plasticity" between osteoblast and adipocyte cells, and it is becoming increasingly clear that a change in BMSC dynamics can result in osteoporosis with age because of an increase in the number of marrow adipocytes at the expense of osteoblasts.[96] The differentiation of the osteoblast/adipocyte bipotential stromal cell into either the osteoblastic or adipocytic phenotype depends on the kinds of signals that it receives, including hormonal, and cytokine stimuli among many others in the marrow microenvironment.[93,97,98] For example, various BMPs and the Wnt pathway induce osteoblast differentiation both independently and synergistically, whereas Dlk-1/Pref-1 and noggin inhibit the same process.[86,99–101] In the same manner, marrow adipogenesis may be stimulated or inhibited, depending on the signals or cues present in the stromal cell milieu. For example, BMP4 and insulin are known inducers of adipocytic differentiation in the periphery and in the marrow, and so also are natural ligands to PPARγ, such as prostaglandin J2 derivatives.[102–104] It is interesting in this context that osteoblasts derived from aging bone marrow of mice display significant reduction in Wnt pathway gene expression and an upregulation in the Wnt antagonist DKK-1.[105]

Aging in mice also results in decreases in osteoblast as well as adipocyte colonies formed from BMSCs in vitro and downregulation of the pro-osteogenic TGF-beta/BMP pathway.[25,106] Increased lipid oxidation with aging has been shown to cause oxidative stress, increased expression of PPARγ, and diminished pro-osteogenic Wnt signaling in the skeleton possibly involving FoxO transcription factors, which also play important roles in maintaining the antioxidant status in aging cells; this however, needs validation in BMSCs.[107] PPARγ 2 controls multiple regulatory pathways of osteoblast differentiation from marrow mesenchymal stem cells and is being proposed as a "master regulator" of osteo/adipogenesis in the marrow.[108]

The role of hormones in the aging marrow is of interest, and elderly women who are well past menopause have low levels of the hormone estrogen.[1] It is well known that estrogen promotes osteoblastic differentiation and inhibits adipocytic differentiation of bone marrow-derived stromal cell lines.[109] It was recently established that loss of estrogens in aged osteoporotic postmenopausal women is associated with significant increase of marrow adipocyte volume, and treatment with an estrogen patch for a one-year time period resulted in significant decreases in marrow adipocyte volume and prevention of adipocyte hyperplasia, as compared to placebo treated controls.[43] In the same cohort of patients, we had earlier demonstrated that estrogen treatment increased bone mineral density.[110] Human mesenchymal stromal cells also express follicle stimulating hormone receptors, and follicle stimulating hormone acts directly on mature bone cells in the skeleton and it is conceivable that direct effects on BMSCs is also possible.[111,112]

The exact aging microenvironment "seen" by BMSCs is not yet certain. Attempts have been made to understand it by employing a "conditioned media" approach in vitro.[92] Recently, serum obtained from young and old donors was added to media and exposed to MSCs in vitro.[113] It was found that the sera of elderly donors was inhibitory to osteoblast but permissive to adipocyte differentiation of these cells; however, the putative factor(s) responsible for these effects need to be determined and will presumably be an active area of investigation in the future.[113]

## Evolving paradigms and conclusions

Age-related osteoporosis is a multifaceted disease that involves a number of distinct tissue pathologies such as decreases in trabecular and cortical bone density, volume, and thickness, and also disorganization of the extracellular matrix. Although the various studies outlined above in humans, animals, and cells have been carried out in recent years, the list is not exhaustive, as new studies are being published. It should also be appreciated that other comorbid conditions, such as diabetes and inflammation, or use of various drugs may impact the pattern and severity of this disease in the aging population. However, what is becoming clear from studies with animal models is that although the aging bone phenotype can be impacted by gender, race, and genetics, the underlying phenotype and cellular and molecular signals seem to be similar. Thus, we observe decrease in bone volume and increase in marrow fat in mice and (wo)men, degeneration of various marrow components and poor quality of the extracellular matrix as a consequence of age, and upregulation of pro-adipogenic and anti-osteogenic cellular signaling pathways, which in turn affect the BMSC commitment of bipotential stromal cells toward adipocytes and away from osteoblasts. These studies also inform us of the usefulness of animal and cellular models of senescence and bone aging in trying to tease out the contributions of various bone cell components in the aging process, and potentially inform us of future targets for drug design and other nonpharmacological interventions.

## Conflicts of interest

The authors declare no conflicts of interest.

## References

1. Riggs, B.L. *et al.* 1982. Changes in bone mineral density of the proximal femur and spine with aging. Differences between the postmenopausal and senile osteoporosis syndromes. *J. Clin. Invest.* **70:** 716–723.
2. Riggs, B.L., S. Khosla & L.J. Melton III. 2001. The type I/type II model for involutional osteoporosis: update and modifications based on new observations. In *Osteoporosis.* 2nd ed. R. Marcus, D. Feldman & J. Kelsey, Eds.: 49–58. Academic Press.
3. Robey, P.G. & A.L. Boskey. 2008. The composition of bone. *Primer on the Metabolic Bone Diseases and Disorders of Mineral Metabolism.* 7th ed. American Society for Bone and Mineral Research. pp. 32–38.
4. Bonewald, L.F. 2008. Osteocytes. *Primer on the Metabolic Bone Diseases and Disorders of Mineral Metabolism.* 7th ed. American Society for Bone and Mineral Research. pp. 32–38.
5. Riggs, B.L., S. Khosla & L.J. Melton III. 2002. Sex steroids and the construction and conservation of the adult skeleton. *Endocr. Rev.* **23:** 279–302.
6. Lee, C.A. & T.A. Einhorn. 2001. The bone organ system: form and function. *Osteoporosis.* 2nd ed. R. Marcus, D. Feldman & J. Kelsey, Eds.: 2–20. Academic Press.
7. Hendrickx, A.G. & W.R. Dukelow. 1995. Reproductive biology. *Nonhuman Primates in Biomedical Research: Biology and Management.* B.T. Bennett, C.R. Abee & R. Henrickson. Eds.: 147–191. Academic press.
8. Black, A. & M.A. Lane. 2002. Nonhuman primate models of skeletal and reproductive aging. *Gerontology* **48:** 72–80.
9. Lees, C.J. & H. Ramsay. 1999. Histomorphometry and bone biomarkers in cynomolgus females: a study in young, mature and old monkeys. *Bone* **24:** 25–28.
10. Jerome, C.P. & P.E. Peterson. 2001. Nonhuman primate models in skeletal research. *Bone* **29:** 1–6.
11. Borsari, V. *et al.* 2007. Sandblasted titanium osteointegration in young, aged and ovariectomized sheep. *Int. J. Artif. Organs* **30:** 163–172.
12. Sachse, A. *et al.* 2005. Osteointegration of hydroxyapatite-titanium implants coated with nonglycosylated recombinant human bone morphogenetic protein-2 (BMP-2) in aged sheep. *Bone* **37:** 699–710.
13. Kettler, A. *et al.* 2007. Are the spines of calf, pig and sheep suitable models for pre-clinical implant tests? *Eur. Spine J.* **16:** 2186–2192.
14. Vacanti, V. *et al.* 2005. Phenotypic changes of adult porcine mesenchymal stem cells induced by prolonged passaging in culture. *J. Cell Physiol.* **205:** 194–201.
15. Lebedinskaia, O.V. *et al.* 2004. Age changes in the number of stromal-precursor cells in the animal bone marrow. *Morfologiia* **126:** 46–49.
16. Huja, S.S., A.M. Rummel & F.M. Beck. 2008. Changes in mechanical properties of bone within the mandibular condyle with age. *J. Morphol.* **269:** 138–143.
17. Sumner, D.R. *et al.* 2003. Aging does not lessen the effectiveness of TGFbeta2-enhanced bone regeneration. *J. Bone Miner. Res.* **18:** 730–736.
18. Tonna, E.A. 1975. Accumulation of lipofuscin (age pigment) in aging skeletal connective tissues as revealed by electron microscopy. *J. Gerontol.* **30:** 3–8.
19. Bikle, D.D. *et al.* 2002. Insulin-like growth factor 1 is required for the anabolic actions of parathyroid hormone on mouse bone. *J. Bone Miner. Res.* **17:** 1570–1578.
20. Cao, J., L. Venton, T. Sakata & B.P. Halloran. 2003. Expression of RANKL and OPG correlates with age-related bone loss in male C57BL/6 mice. *J. Bone Miner. Res.* **18:** 270–277.
21. Watanabe, K., A. Hishiya. 2005. Mouse models of senile osteoporosis. *Mol. Aspects Med.* **26:** 221–231.
22. Ferguson, V.L., R.A. Ayers, T.A. Bateman & S.J. Simske. 2003. Bone development and age related bone loss in male C57BL/6 mice. *Bone* **33:** 387–398.
23. Cooper, D.M.L. *et al.* 2007. Age dependent change in the 3D stucture of cortical porosity at the human femoral midshaft. *Bone* **40:** 957–965.
24. Bergman, R.J. *et al.* 1996. Age related changes in the osteogenic stem cells in mice. *J. Bone Miner. Res.* **11:** 568–577.
25. Zhang, W. *et al.* 2008. Age-related changes in the osteogenic differentiation potential of mouse bone marrow stromal cells. *J. Bone Miner. Res.* **23:** 1118–1128.
26. Morrison, S.J. *et al.* 1996. The aging of hematopoietic stem cells. *Nat. Med.* **2:** 1011–1016.
27. LeMaoult, J. *et al.* 1999. Cellular basis of B cell clonal populations in old mice. *J. Immunol.* **162:** 6384–6391.
28. Ghia, P., F. Melchers & A.G. Rolink. 2000. Age dependent changes in B lymphocyte development in man and mouse. *Exp.Gerontol.* **35:** 159–165.
29. Kerschan-Schindl, K. 2008. Serum levels of receptor activator of nuclear factor kappaB ligand (RANKL) in healthy women and men. *Exp. Clin. Endocrinol. Diabetes* **116:** 491–495.
30. Pietschmann, P. *et al.* 2009. Osteoporosis: an age related and gender specific disease: a mini review. *Gerontology* **55:** 3–12.
31. Iida, H. & S. Fukuda. 2002. Age-related changes in bone mineral density, cross sectional area and strength at different skeletal sites in male rats. *J. Vet. Med. Sci.* **64:** 29–34.
32. Fukuda, S. & H. Iida. 2004. Age-related changes in bone mineral density, cross sectional area and strength of long bones in the hind limbs and first lumbar vertebra in female Wistar rats. *J. Vet. Med. Sci.* **66:** 755–756.

33. Duque, G. *et al.* 2009. Age-related bone loss in the LOU/c rat model of healthy aging. *Exp. Gerentol.* **44:** 183–189.

34. Sabsovich, I. *et al.* 2008. Bone microstructure and its genetic variability in 12 inbred strains: μCT study and *in silico* genome scan. *Bone* **42:** 439–451.

35. Syed, F. & S. Khosla. 2005. Mechanisms of sex steroid effects on bone. *Biochem. Biophys. Res. Commun.* **328:** 688–696.

36. Priemel, M. *et al.* 2002. Osteopenic mice: animal models of the aging skeleton. *J. Musculoskelet Neuronal Interact.* **2:** 212–218.

37. Falahati-Nini, A. *et al.* 2000. Relative contributions of testosterone and estrogen in regulating bone resorption and formation in normal elderly men. *J. Clin. Invest.* **106:** 1553–1560.

38. McCauley, L.K., T.F. Tozum & T.J. Rosol. 2002. Estrogen receptors in skeletal metabolism: lessons from genetically modified models of receptor function. *Crit. Rev. Eukaryot. Gene Expr.* **12:** 89–100.

39. Almeida, M. *et al.* 2007. Skeletal involution by age-associated oxidative stress and its acceleration by loss of sex steroids. *J. Biol. Chem.* **282:** 27285–27297.

40. Duque, G. *et al.* 2002. Estrogens (E2) regulate expression and response of 1,25-dihydroxyvitamin D3 receptors in bone cells: changes with aging and hormone deprivation. *Biochem. Biophys. Res. Commun.* **299:** 446–454.

41. Dunnill, M.S., J.A. Anderson & R. Whitehead. 1967. Quantitative histological studies on age changes in bone. *J. Path. Bact.* **94:** 275–291.

42. Meunier, P., J. Aaron, C. Edouard & A. Vignon. 1971. Osteoporosis and the replacement of cell populations of the marrow by adipose tissue. *Clin. Orthop. Rel. Res.* **80:** 147–154.

43. Syed, F.A. *et al.* 2008. Effects of estrogen therapy on bone marrow adipocytes in postmenopausal osteoporotic women. *Osteopor. Int.* **19:** 1323–1330.

44. Syed, F. *et al.* 2010. Effects of chronic estrogen treatment on modulating age-related bone loss in female mice. *J. Bone Miner. Res.* Published online.

45. Takeda, T. *et al.* 1981. A new murine model of accelerated senescence. *Mech. Ageing Dev.* **17:** 183–194.

46. Matsushita, M. *et al.* 1986. Age-related changes in bone mass in the senescence accelerated mouse (SAM) SAM-R/3 and SAM/6 as new models for osteoporosis. *Am. J. Pathol.* **215:** 276–283.

47. Jilka, R.L. *et al.* 1996. Linkage of decreased bone mass with impaired osteoblastogenesis in a murine model of accelerated senescence. *J. Clin. Invest.* **97:** 1732–1740.

48. Silva, M.J., M.D. Brodt, M. Ko & Y. Abu-Amer. 2005. Impaired marrow osteogenesis is associated with reduced endocortical bone formation but does not impair periosteal bone formation in long bones of SAMP6 mice. *J. Bone Miner. Res.* **20:** 419–427.

49. Kasai, S. *et al.* 2004. Consistency of low bone density across bone sites in SAMP6 laboratory mice. *J. Bone Miner. Metab.* **22:** 207–214.

50. Okamoto, Y. *et al.* 1995. Femoral peak bone mass and osteoclast number in an animal model of age-related spontaneous osteopenia. *Anat. Rec.* **242:** 21–28.

51. Kajnekova, O. *et al.* 1997. Increased adipogenesis and myelopoiesis in the bone marrow of SAMP6, a murine model of defective osteoblastogenesis and low turnover osteopenia. *J. Bone Miner. Res.* **12:** 1772–1779.

52. Tohjima, E. *et al.* 2003. Decreased AP-1 activity and interleukin-11 expression by bone marrow stromal cells may be associated with impaired bone formation in aged mice. *J. Bone Miner. Res.* **18:** 1461–1470.

53. Clement-Lacroix, P. *et al.* 2005. Lrp5-independent activation of Wnt signaling by lithium chloride increases bone formation and bone mass in mice. *Proc. Natl. Acad. Sci. U.S.A.* **102:** 17406–17411.

54. Nakanishi, R. *et al.* 2006. Secreted frizzled-related protein 4 is a negative regulator of peak BMD in SAMP6 mice. *J. Bone Miner. Res.* **21:** 1713–1721.

55. Kuro-o, M. *et al.* 1997. Mutation of the mouse klotho gene leads to a syndrome resembling aging. *Nature* **390:** 45–51.

56. Kuro-o, M. 2009. Klotho and aging. *Biochim. Biophys. Acta.* **1790:** 1049–1058.

57. Suzuki, H. *et al.* 2005. Histological evidence of the altered distribution of osteocytes and bone matrix synthesis in klotho-deficient mice. *Arch. Histol. Cytol.* **68:** 371–381.

58. Suzuki, H. *et al.* 2008. Histological and elemental analysis of impaired bone mineralization in klotho-deficient mice. *J. Anat.* **212:** 275–285.

59. Kawaguchi, H. *et al.* 1999. Independent impairment of osteoblast and osteoclast differentiation in klotho mouse exhibiting low turnover osteopenia. *J. Clin. Invest.* **104:** 229–237.

60. Kuro-o, M. 2008. Klotho as a regulator of oxidative stress and senescence. *Biol. Chem.* **389:** 233–241.

61. Suzuki, H. *et al.* 2008. Involvement of the klotho protein in dentin formation and mineralization. *Anat. Rec.* **291:** 183–190.

62. Liu, H. *et al.* 2007. Augmented Wnt signaling in a mammalian model of accelerated aging. *Science* **317:** 803–806.

63. Verstraeten, V.L. *et al.* 2007. The nuclear envelope, a key structure in cellular integrity and gene expression. *Curr. Med. Chem.* **14:** 1231–1248.

64. Mounkes, L.C. *et al.* 2003. A progeroid syndrome in mice is caused by defects in A-type lamins. *Nature* **423:** 298–301.

65. Akter, R. *et al.* 2009. Effect of lamin A/C knockdown on osteoblast differentiation and function. *J Bone Miner. Res.* **24:** 283–293.

66. Rivas, D., W. Li, R. Akter, *et al.* 2009. Accelerated features of age-related bone loss in zmpste24 metalloproteinase-deficient mice. *J. Gerontol. A Biol. Sci. Med. Sci.* **64:** 1015–1024.

67. Le Gall, J.Y. & R. Ardaillou. 2009. The biology of aging. *Bull. Acad. Natl. Med.* **193:** 365–402.

68. Allori, A.C., A.M. Sailon & S.M. Warren. 2008. Biological basis of bone formation, remodeling, and repair-part II: extracellular matrix. *Tissue Eng. Part B Rev.* **14:** 275–283.

69. Wang, C.J. *et al.* 2008. Trabecular bone deterioration in col9a1+/− mice associated with enlarged osteoclasts adhered to collagen IX deficient bone. *J. Bone Miner. Res.* **23:** 837–849.

70. Kato, N. *et al.* 2007. Osteopontin deficiency enhances anabolic action of EP4 agonist at a sub-optimal dose in bone. *J. Endocrinol.* **193:** 171–182.

71. Ikeda, T. *et al.* 1995. Age-related reduction in bone matrix protein mRNA expression in rat bone tissues: application of histomorphometry to *in situ* hybridization. *Bone* **16**: 17–23.

72. Young, M.F., Y. Bi, L. Ameye & X.D. Chen. 2002. Biglycan knockout mice: new models for musculoskeletal diseases. *Glycoconj. J.* **19**: 257–262.

73. Chen, X.D. *et al.* 2002. Age-related osteoporosis in biglycan-deficient mice is related to defects in bone marrow stromal cells. *J. Bone Miner. Res.* **17**: 331–340.

74. McBeath, R. *et al.* 2004. Cell shape, cytoskeletal tension and RhoA regulate stem cell lineage commitment. *Dev. Cell* **6**: 483–495.

75. Windahl, S.H. *et al.* 2001. Female estrogen receptor beta−/− mice are partially protected against age-related trabecular bone loss. *J. Bone Miner. Res.* **16**: 1388–1398.

76. Juttner, K.V. & M.J. Perry. 2007. High-dose estrogen-induced osteogenesis is decreased in aged RUNX2 (+/−) mice. *Bone* **41**: 25–32.

77. Tsuji, K., T. Komori & M. Noda. 2004. Aged mice require full transcription factor, Runx2/Cbfa1, gene dosage for cancellous bone regeneration after bone marrow ablation. *J. Bone Miner. Res.* **19**: 1481–1489.

78. Ducy, P. *et al.* 1999. A Cbfa-1 dependent genetic pathway controls bone formation beyond embryonic development. *Genes Dev.* **13**: 1025–1036.

79. Akune, T. *et al.* 2004. PPARgamma insufficiency enhances osteogenesis through osteoblast formation from bone marrow progenitors. *J. Clin. Invest.* **113**: 846–855.

80. Wu, X.B. *et al.* 2005. Impaired osteoblastic differentiation, reduced bone formation, and severe osteoporosis in noggin-overexpressing mice. *J. Clin. Invest.* **112**: 924–934.

81. Baylink, D., KH-W. Lau & S. Mohan. 2007. The role of IGF system in the rise and fall in bone density with age. *J. Musculoskelet. Neuronal Interact.* **7**: 304–305.

82. Karsenty, G. & F. Elefteriou. 2008. Neuronal regulation of bone remodeling. *Primer on the Metabolic Bone Diseases and Disorders of Mineral Metabolism,* 7th ed.: 56–60. American Society for Bone and Mineral Research.

83. Ducy, P. *et al.* 2000. Le6ptin inhibits bone formation through a hypothalamic relay: a central control of bone mass. *Cell* **100**: 197–207.

84. Elefteriou, F. *et al.* 2005. Leptin regulation of bone resorption by the sympathetic nervous system and CART. *Nature* **434**: 514–520.

85. Bouxsein, M.L. 2009. Mice lacking beta-adrenergic receptors have increased bone mass but are not protected from deleterious skeletal effects of ovariectomy. *Endocrinology* **150**: 144–152.

86. Idris, A.I. *et al.* Cannabinoid receptor type 1 protects against age-related osteoporosis by regulating osteoblast and adipocyte differentiation in marrow stromal cells. *Cell Metab.* **10**: 139–147.

87. Hamrick, M.W. *et al.* 2006. Age-related loss of muscle mass and bone strength in mice is associated with a decline in physical activity and serum leptin. *Bone* **39**: 845–853.

88. Hamrick, M.W. *et al.* 2004. Leptin deficiency produces contrasting phenotypes in bones of the limb and spine. *Bone* **34**: 376–383.

89. Teitelbaum, S.L. 2000. Bone resorption by osteoclasts. *Science* **289**: 1504–1508.

90. Owen, M. 1988. Marrow stromal cells. *J. Cell Sci.* **10**: 63–76.

91. Chamberlain, G., J. Fox, B. Ashton & J. Middleton. 2007. Concise review: mesenchymal stem cells: their phenotype, differentiation capacity, immunological features and potential for homing. *Stem Cells* **25**: 2739–2749.

92. Bellantuono, I., A. Aldahmash & M. Kassem. 2009. Aging of marrow stromal (skeletal) stem cells and their contribution to age-related bone loss. *Biochim. Biophys. Acta.* **1792**: 364–370.

93. Gimble, J.M. *et al.* 2006. Playing with bone and fat. *J. Cell. Biochem.* **98**: 251–266.

94. Nakashima, K. *et al.* 2002. The novel zinc-finger containing transcription factor osterix is required for osteoblast differentiation and bone formation. *Cell* **108**: 17–29.

95. Lazar, M.A. 2005. PPAR gamma, 10 years later. *Biochimie.* **87**: 9–13.

96. Beresford, J.N. *et al.* 1992. Evidence for an inverse relationship between the differentiation of adipocytic and osteogenic cells in rat marrow stromal cell cultures. *J. Cell. Sci.* **102**: 341–351.

97. Kassem, M., B.M. Abdallah & H. Saeed. 2008. Osteoblastic cells: differentiation and trans-differentiation. *Arch. Biochem. Biophys.* **473**: 183–187.

98. Kang, Q. *et al.* 2009. A comprehensive analysis of the dual roles of BMPs in regulating adipogenic and osteogenic differentiation of mesenchymal progenitor cells. *Stem Cells Dev.* **18**: 545–559.

99. Gaur, T. *et al.* 2005. Canonical WNT signaling promotes osteogenesis by directly stimulating Runx2 gene expression. *J. Biol. Chem.* **280**: 33132–33140.

100. Abdallah, B.M. *et al.* 2004. Regulation of human skeletal stem cells differentiation by Dlk1/Pref-1. *J. Bone Miner. Res.* **19**: 841–852.

101. Rifas, L. 2007. The role of noggin in human mesenchymal stem cell differentiation. *J. Cell. Biochem.* **100**: 824–834.

102. Tang, Q.Q., T.C. Otto & M.D. Lane. 2004. Commitment of C3H10T1/2 pluripotent stem cells to the adipocyte lineage. *Proc. Natl. Acad. Sci. USA* **101**: 9607–9611.

103. Ding, J., K. Nagai, J.T. Woo. 2003. Insulin-dependent adipogenesis in stromal ST2 cells derived from murine bone marrow. *Biosci. Biotechnol. Biochem.* **67**: 314–321.

104. Nosjean, O. & J.A. Boutin. 2002. Natural ligands of PPARgamma: are prostaglandin J(2) derivatives really playing the part. *Cell Signal.* **14**: 573–583.

105. Rauner, M., W. Sipos & P. Pietschmann. 2008. Age-dependent Wnt gene expression in bone and during the course of osteoblast differentiation. *Age (Dordr).* **30**: 273–282.

106. Moerman, E.J. *et al.* 2004. Aging activates adipogenic and suppresses osteogenic programs in mesenchymal marrow stroma/stem cells: the role of PPAR-gamma2 transcription

factor and TGF-beta/BMP signaling pathways. *Aging Cell.* **3:** 379–389.

107. Almeida, M. *et al.* 2009. Increased lipid oxidation causes oxidative stress, increased peroxisome proliferator-activated receptor gamma expression, and diminished pro-osteogenic Wnt signaling in the skeleton. *J. Biol. Chem.* **284:** 27438–27448.

108. Shockley, K.R. *et al.* 2009. PPARgamma2 nuclear receptor controls multiple regulatory pathways of osteoblast differentiation from marrow mesenchymal stem cells. *J. Cell. Biochem.* **106:** 232–246.

109. Okazaki, R. *et al.* 2002. Estrogen promotes early osteoblast differentiation in mouse bone marrow stromal cell lines that express estrogen receptor alpha or beta. *Endocrinology* **143:** 2349–2356.

110. Lufkin, E.G. *et al.* 1992. Treatment of postmenopausal osteoporosis with transdermal estrogen. *Ann. Intern. Med.* **117:** 1–9.

111. Sun, L. *et al.* 2006. FSH directly regulates bone mass. *Cell.* **125:** 247–260.

112. Imam, A. *et al.* 2009. Role of pituitary-bone axis in skeletal pathophysiology. *Curr. Opin. Endocrinol. Diabetes Obes.* In press.

113. Abdallah, B.M. *et al.* 2006. Inhibition of osteoblast differentiation of mesenchymal stem cells by sera obtained from aged females. *Bone* **39:** 181–188.

Ann. N.Y. Acad. Sci. ISSN 0077-8923

ANNALS OF THE NEW YORK ACADEMY OF SCIENCES
Issue: *Molecular and Integrative Physiology of the Musculoskeletal System*

# The crossover of bisphosphonates to cancer therapy

Merry Sun, Jameel Iqbal, Sukhjeet Singh, Li Sun, and Mone Zaidi

The Mount Sinai Bone Program, Department of Medicine, Mount Sinai School of Medicine, New York, New York

Address for correspondence: Mone Zaidi, M.D., Ph.D., The Mount Sinai Bone Program, Department of Medicine, Mount Sinai School of Medicine, New York, NY 10029. Mone.zaidi@mssm.edu

Bisphosphonates form a class of drugs commonly used to treat disorders of osteoclastic bone resorption, including osteoporosis, Paget's disease of the bone, rheumatoid arthritis, and bone metastases. Although long established as the therapy of choice to treat such disorders, bisphosphonates' potential in treating cancer is garnering interest. Bisphosphonates have been demonstrated to inhibit tumor growth and metastasis, induce apoptosis in tumor cells, and encourage immune reactions against tumor cells. Current applications of bisphosphonates in cancer treatment include their use to treat skeletal metastases and as an adjuvant to endocrine therapy. This review explores bisphosphonates' current clinical utility and potential as a crossover cancer therapy.

Keywords: bone metastases; bisphosphonates; cancer therapy

## Introduction

Bisphosphonates are primarily known for their ability to control excessive bone resorption in diseases such as osteoporosis, Paget's disease of the bone, rheumatoid arthritis (RA), periodontal disease, and osteolytic bone metastases. Bisphosphonates inhibit bone resorption by interrupting the mevalonate pathway and thereby blocking the prenylation and activation of GTPases necessary to maintain osteoclastic function and survival. They have an affinity for calcium, which is a major component of skeletal hydroxyapatite. Therefore, the effects of bisphosphonates are most clearly seen in bone.

Bisphosphonates have been used extensively and successfully to treat osteolytic bone metastases, which are especially common in patients with breast, prostate, renal cell, and lung cancers. Their success in the therapy of metastases and other skeletal-related events associated with cancer, including treatment-induced bone loss, has been the subject of recent investigations.

Newer studies, however, demonstrate that certain bisphosphonates, such as zoledronic acid, may have direct effects on the growth of tumor cells. For example, zoledronic acid decreases the expression of vascular endothelial growth factor (VEGF), a cytokine needed for angiogenesis. Additionally, bisphosphonates, most notably zoledronic acid and pamidronate, display other antitumor actions such as the induction of apoptosis.[1,2] This review summarizes the anti-osteoclastic actions of bisphosphonates and then focuses on their recently described antitumor effects.

## Bisphosphonates and osteoporosis

*Discovery and structure–activity relationships*
Bisphosphonates were discovered in the 1960s; however, chemists had synthesized similar molecules since the late 19th century. Bisphosphonates are structural and functional analogs of pyrophosphates. Both bisphosphonates and pyrophosphates have a strong affinity for divalent cations, including calcium. The clinical utility of bisphosphonates developed from the need to duplicate pyrophosphates' ability to bind calcium phosphate, thereby inhibiting calcification.

The P-C-P backbone of bisphosphonates is similar to the P-O-P backbone of pyrophosphates.[3] This backbone stabilizes the molecule, resists phosphatase activity, and encourages binding to the mineralized bone matrix.[4] Bisphosphonates are classified by their unique chemical R groups and corresponding mechanism of action; there are

doi: 10.1111/j.1749-6632.2010.05812.x

two general groups: the nitrogen (N)-containing bisphosphonates and the non-N-containing bisphosphonates. The N-containing bisphosphonates are alendronate, ibandronate, incadronate, minodronate, neridronate, pamidronate, risedronate, tiludronate, and zoledronate. Currently, four of these—namely, alendronate, risedronate, ibandronate, and zoledronic acid—are approved for the prevention and treatment of osteoporosis.

N-containing bisphosphonates are generally more potent than their non-N counterparts. The relative efficacy of the N-containing compounds depends on the side chain that directly interacts with farnesyl pyrophosphate synthase (FPPS) to interrupt the mevalonate pathway.[5] During osteoclastic bone resorption, bisphosphonates are pinocytosed; within the osteoclast, they inhibit FPPS, thereby preventing the synthesis of geranygeranyl diphosphate (GGPP). FPPS and GGPP play a large role in prenylation of small GTPases. Ras proteins with a cysteine–aliphatic–aliphatic–arbitrary ("CaaX box") undergo prenylation by an FPPS molecule or a GGPP molecule, with the aid of farnesyl transferase or geranylgeranyl transferase, respectively. The small GTPase Ras and its relatives are involved in processes, including cell cycle regulation, nuclear import, apoptosis, and differentiation.[6] Prenylation of oncogenic forms of Ras has been implicated in malignant transformation of cells.[7]

This metabolic block results most prominently in the inhibition of prenylation of small GTPases needed for motility and secretory activity.[5,8] In contrast, non-N-containing bisphosphonates do not inhibit FPPS and, instead, are thought to work by a different mechanism. They combine with aminoacyladenylate to form nonhydrolyzable, toxic analogs of ATP.[4,5,8,9]

### Use of bisphosphonates in osteoporosis treatment

The absorption of oral bisphosphonates is very low, in the region of 0.6%. Nonetheless, pivotal, large, well-controlled clinical trials have shown that oral therapy reduces the risk of vertebral fractures, and in certain instances, nonvertebral and hip fractures.[10] Overall, bisphosphonates, when used to prevent or treat osteoporosis, increase bone density at all sites and reduce bone turnover to premenopausal levels. Newer bisphosphonates, including intravenous

ibandronate and zoledronic acid, obviously provide more robust bioavailability when compared with oral therapies. Approximately 50–75% of a dose of intravenously administered bisphosphonates binds to the bone matrix. The rest is excreted by the kidney.[4]

## Bisphosphonates and bone metastases

### Bone microenvironment, metastases, and bisphosphonate

Bone metastases are common in breast, lung, and prostate cancers, among other cancers. Metastases can be osteoblastic, osteolytic, or mixed. Approximately 70% of patients who die from breast or prostate cancer have bone metastases. Likewise, ~40% of those who die from thyroid, kidney, or lung cancers have bone metastases. The skeletal system is a common migration site for metastatic cancer cells.[11] Mineralized bone contains cytokines, chemokines, and growth factors that create a fertile environment for tumor cells. Osteoclastic resorption releases these growth factors, thereby promoting metastatic tumor cell growth.

Furthermore, metaphyseal bone has an extensive system of blood vessels adapted for the easy passage of cells in and out of the marrow. This property is ideal for the spread of metastatic cells. Instead of capillary beds, the metaphyseal vasculature is composed largely of sinusoids. The sinusoid lumen has a diameter far larger than what is needed to accommodate a cancer cell. Sinusoids are also highly permeable and, in the metaphyseal bone, are located only a few microns away from the spongy bone. Accordingly, metastases tend to reside in marrow-rich, well-vascularized areas such as the spinal vertebrae, ribs, and ends of long bones.

Metaphyses are also rich in bone marrow; red bone marrow is, in turn, rich in hematopoietic, mesenchymal, and also stromal cells. These latter cell types are progenitors of a variety of cell lineages such as osteoblasts and adipocytes that secrete different tumor-attractive cytokines, chemokines, and growth factors.[11] Finally, the bone microenvironment contains adhesion proteins that encourage cancer cell colonization. Stromal and mesenchymal cells, which can differentiate into adipocytes, fibroblasts, chodrocytes, and osteoblasts, express vascular cell adhesion molecules.

Osteolytic metastases frequently lead to fracture and osteoporosis. N-containing bisphosphonates taken orally or intravenously alleviate symptoms of osteolytic metastases and other cancer-treatment-induced bone loss. Specifically, they have been shown to decrease pathologic fractures, bone pain, hypercalcemia, spinal compression, and the need for palliative radiation therapy to bone metastases.[12] These effects are largely attributed to the potent antiresorptive effects of bisphosphonates, wherein through inhibition of bone resorption, they make the bone microenvironment less hospitable to tumor cells. It is possible, however, that bisphosphonates have direct antitumor cell effects (see below).

## Bisphosphonates for the prevention of cancer treatment–induced bone loss

Systemic therapies, such as chemotherapy and endocrine therapy, are commonly used to prevent cancer recurrence and/or limit metastasis growth in bone.[12] Drugs used in endocrine therapy for breast cancer include aromatase inhibitors such as anastrozole and letrozole; gonadotropin-releasing hormone (GnRH) agonists such as goserelin; luteinizing hormone-releasing hormone agonists; and selective estrogen receptor modulators such as tamoxifen.[13] In the treatment of breast cancer, these drugs serve to suppress estrogen production, thereby halting estrogen-sensitive tumor growth. However, treatment-induced hypoestrogenemia increases osteoclastic bone resorption leading to acute, severe bone loss. The intravenous bisphosphonate zoledronic acid has demonstrated effectiveness in decreasing bone mineral density losses in breast cancer patients undergoing therapy with aromatase inhibitors and tamoxifen.[14]

Although bisphosphonates improve quality of life for patients by preventing dramatic losses in bone mineral density, which, in turn, alleviates chronic pain, loss of mobility, and fractures that are associated with cancer treatment–induced bone loss, they do not significantly impact survival time. Gnant and colleagues[15] investigated the efficacy of zoledronic acid as an adjuvant for endocrine therapy in breast cancer patients. Patients were randomly assigned to receive goserelin plus tamoxifen or anastrozole with or without zoledronic acid for three years. At a median follow-up point of 47.8 months, disease progression risk for patients undergoing therapy with zoledronic acid were 3.2 absolute percentage points lower and 36 relative percentage points lower than disease progression risk for patients undergoing endocrine therapy without zoledronic acid. However, no significant difference in disease-free survival rates was noted between the endocrine therapy-only groups, the anastrozole, and the tamoxifen groups.[15] Likewise, Mauri and colleagues[16] examined the efficacy of bisphosphonates as an adjuvant therapy in altering the course of the disease for breast cancer in a meta-analysis of 13 clinical trials. Only trials that studied direct effect of bisphosphonate therapy on the course of patients' breast cancer were considered; trials that concentrated on bisphosphonates' efficacy in alleviating cancer-treatment-induced bone disease were not considered.[16] Six trials tested zoledronic acid, two tested pamidronate, one tested risedronate, and four tested clodronate. All trials were randomized and used only patients with primary breast cancer to undergo either adjuvant therapy alone or with additional bisphosphonate therapy. There was no significant benefit to the use of bisphosphonates in limiting recurrences or bone metastases. However, a positive but statistically insignificant trend favored bisphosphonate use in reducing deaths. It was also noted that zoledronic acid may be effective in decreasing disease recurrences.[16]

In men with prostate cancer, endocrine therapies include androgen deprivation. Androgen-deprivation therapies include surgical orchiectomy, as well as drugs such as GnRH agonists and antagonists, and anti-androgens. These therapies decrease circulating androgens, especially testosterone, which augments tumor growth. Androgen-deprivation therapy causes a dramatic loss of bone mass with a resulting high fracture risk. In these instances, alendronate taken orally has only limited efficacy,[17] whereas the more potent intravenous bisphosphonates have been found to improve bone mass significantly. Clinical trials with zoledronic acid have shown reduced bone pain and skeletal complications in prostate cancer patients with bone metastases.[17] Zoledronic acid has also demonstrated efficacy in preventing cancer treatment–induced bone loss in patients treated with androgen deprivation, improving bone mineral density.[17] Bisphosphonate therapy may decrease or delay incidence of skeleton-related events, such as bone metastases, at later stages.[18,19]

## Effects of bisphosphonates on tumor growth

### Direct effects on tumor cells

There are various mechanisms by which bisphosphonates are thought to affect tumor cells directly. *In vitro*, the drugs inhibit tumor cell adhesion, angiogenesis, invasion, proliferation, and survival. For example, zoledronic acid and pamidronate both induce apoptosis.[1,20] The bisphosphonates can induce apoptosis in a variety of tumors: *in vitro*, they induce apoptosis in breast, prostate, ovarian, melanoma, colon, osteosarcoma, pancreatic, myeloma, and leukemia cells. One mechanism of induction of apoptosis is mitochondrial-dependent; pamidronate and zoledronic acid-mediated apoptosis in breast and prostate cancer occurs through downregulation of the anti-apoptotic factor Bcl-2, release of mitochondrial cytochrome *c*, and ensuing caspase activation.[1,21] Simultaneously, disruption of the mevalonate pathway results in the inability to prenylate small GTPases such as Ras that are frequently needed for activation of the anti-apoptotic mitogen-activated protein kinase and Akt signaling pathways.[9]

Bisphosphonates inhibit the adhesion of breast and prostate cancer cells when cortical bone slices or tumor cells are pretreated with a drug.[1,9] It is likely that this effect occurs via interruption of small GTPase prenylation; small GTPases are needed for the activation of integrins in cell adhesion.[9] For example, at low concentrations, alendronate and zolendronate inhibit the adhesion and motility of ovarian and breast cancer cells, respectively, by decreasing geranylgeranylation of RhoA.[9]

Regarding angiogenesis and tumor migration, zoledronic acid has been shown to decrease levels of the tumor angiogenesis cytokine VEGF.[2] The inhibition of tumor migration likely occurs through bisphosphonates' ability to inhibit the zinc-dependent proteolytic activity of matrix metalloproteinases (MMPs) at high concentrations ($10^{-4}$ M). It should be noted that *in vivo* concentrations that effectively inhibit bone resorption and possibly affect tumor cell adhesion and invasion may not be high enough to affect MMP-assisted migration. Additionally, zoledronic acid can inhibit breast cancer migration through de-

creasing the surface expression of chemokines receptors such as CXC chemokine receptor 4 (CXCR4).[9]

### Effect on isolated tumor cells in bone marrow

Bone marrow micrometastasis, or the presence of isolated tumor cells (ITCs) in the bone marrow, significantly shortens overall survival, as well as disease-free survival and distant disease-free intervals in breast cancer patients, while increasing risk for recurrence.[22] ITCs thrive in the bone marrow microenvironment, as they can adapt and grow for years before proliferating and disseminating.[22–24] Although dormant in the bone marrow, ITCs remain in the $G_0$ cell cycle, rendering cytostatic treatments for ITCs ineffective. Zoledronic acid, a cell cycle-independent drug, has demonstrated an antineoplastic effect against ITCs and has therefore shown potential in increasing survival and decreasing recurrence in breast cancer patients.[23,24]

Rack and colleagues[24] conducted a nonrandomized phase II clinical trial to study the effect of zoledronic acid on ITCs in the bone marrow in breast cancer patients. In a study of 31 women with primary breast cancer that showed persistent ITCs in their bone marrow, zoledronic acid was given for six months. They found that 87% of the patients in the zolendronate group had eradication of ITCs compared with 73% of the patients in the control group.[24] This successful eradication of ITCs by zoledronic acid is comparable to the clinical effect of the monoclonal antibodies edrecolomab and trastuzumab.[24–26]

Bisphosphonates have not only demonstrated good efficacy in the treatment of bone metastases, but have also shown inhibition of intracellular pathways resulting in direct antitumour activity *in vitro*.[24,27,28] Breast cancer patients receiving bisphosphonates as part of their adjuvant systemic treatment have prolonged survival, pointing also to a clinical benefit of bisphosphonates in the primary setting.[15,24,29–32] Although the mechanism of a therapeutic intervention using bisphosphonates is largely unknown, low expression of proliferation markers, such as Ki-67 and p120, in bisphosphonate-treated ITCs indicates that most of these cells are $G_0$ phase dormant state.[33] Furthermore, the action of bisphosphonates on ITCs may be also based on the prevention of tumor cell adhesion, induction of apoptosis, antagonism of growth factors, and antiangiogenic effects.[24–26]

## Immunomodulatory effects of bisphosphonates

N-containing bisphosphonates have shown potential in stimulating T cells against tumor cells. T cells with the $\gamma\delta$ T cell receptor (TCR) comprise 2–5% of CD3$^+$ T cells in the peripheral blood. Elsewhere in the body, such as the intestine, cells with the $\gamma\delta$ TCR constitute a major subset of T cells. $\gamma\delta$ T cells and $\alpha\beta$ T cells, the more common variety of T cells, have similar effects when activated; the mechanisms of activation are the key difference between the two varieties.[34,35]

Uniquely, $\gamma\delta$ T cells can recognize isopentenylpyrophosphate (IPP), an intermediate of the mevalonate pathway.[35,36] The accumulation of metabolites of the mevalonate system is a powerful danger signal that activates $\gamma\delta$ T cells. Zoledronic acid induces formation of a pro-apoptotic ATP analog, 1-adenosin-5'-yl ester 3-(3-methylbut-3-enyl) ester triphosphoric acid (Apppl), in rabbit osteoclasts *in vivo* due to inhibition of FPPS and the ensuing accumulation of IPP.[37,38] Clodronate induces formation of a similar ATP analog, adenosine 5'($\beta$, $\gamma$-dichloromethylene) triphosphate (AppCCl$_2$p).[37] Apppl was synthesized in rabbit osteoclasts after a single dose of 100 $\mu$g/kg, which is approximately equivalent to a 4 mg single dose in humans, a clinically relevant dose. Apppl inhibits mitochondrial adenine nucleotide translocase, which plays a role in mitochondrial permeability; mitochondrial permeability greatly affects the release of apoptogenic proteins such as cytochrome *c*.

Tumor cells circumvent the buildup of Apppl/AppCCl$_2$p and other metabolites with hydroxylmethylglutaryl-CoA reductase (HMGR). HMGR, a rate-limiting enzyme in the mevalonate pathway, prevents buildup of mevalonate metabolites. Bisphosphonates cause mevalonate disruption and an ensuing overregulation of HMGR. This causes the buildup of mevalonate metabolites that can result in immune recognition by $\gamma\delta$ T cells of breast carcinoma and lymphoma cell lines.[36]

It has been suggested that activated T cells in the arthritic lesion were responsible for osteoclast activation and focal bone destruction, whereas recombination activating genein knockout mice that lack mature lymphocytes develop bone erosions.[39,40] Osteoclasts are important targets for therapeutic interventions to block focal bone erosion in inflammatory arthritis. Data from other animal models of inflammatory arthritis showing that bisphophonates prevent focal bone resorption lend support to this hypothesis.[39,41] However, in patients with RA, treatment with antiresorptive therapies, such as bisphosphonates alone, have not prevented the progression of focal bone erosions.[42] This lack of efficacy may be due to a limited ability to concentrate the drug at sites of inflammation.[39]

## Conclusion

Bisphosphonates are used widely in the treatment of osteolytic bone disorders. Their utility may, however, potentially extend to treatment of primary tumors based on an extensive evidence of their efficacy *in vivo* in limiting tumor adhesion, angiogenesis, metastasis, proliferation, and survival. However, these effects must be carefully investigated further, both *in vitro* and clinically. Nevertheless, it does appear from recent studies that bisphosphonates may indeed be useful in cancer therapeutics.

## Conflicts of interest

The authors declare no conflicts of interest.

## References

1. Senaratne, S.G., G. Pirianov, J.L. Mansi, *et al.* 2000. Bisphosphonates induce apoptosis in human breast cancer cell lines. *Br. J. Cancer* **82:** 1459–1468.
2. Lipton, A. 2008. Emerging role of bisphosphonates in the clinic–antitumor activity and prevention of metastasis to bone. *Cancer Treat. Rev.* **34**(Suppl. 1): S25–S30.
3. Fleisch, H. 2002. Development of bisphosphonates. *Breast Cancer Res.* **4:** 30–34.
4. Coleman, R.E. 2008. Risks and benefits of bisphosphonates. *Br. J. Cancer* **98:** 1736–1740.
5. van Beek, E.R. *et al.* 2003. Differentiating the mechanisms of antiresorptive action of nitrogen containing bisphosphonates. *Bone* **33:** 805–811.
6. Pechlivanis, M. & J. Kuhlmann. 2006. Hydrophobic modifications of Ras proteins by isoprenoid groups and fatty acids—more than just membrane anchoring. *Biochim. Biophys. Acta* **1764:** 1914–1931.
7. Zhang, F.L. & P.J. Casey. 1996. Protein prenylation: molecular mechanisms and functional consequences. *Annu. Rev. Biochem.* **65:** 241–269.
8. Roelofs, A.J., K. Thompson, S. Gordon & M.J. Rogers. 2006. Molecular mechanisms of action of bisphosphonates: current status. *Clin. Cancer Res.* **12:** 6222s–6230s.
9. Clezardin, P., F.H. Ebetino & P.G. Fournier. 2005. Bisphosphonates and cancer-induced bone disease: beyond their antiresorptive activity. *Cancer Res.* **65:** 4971–4974.

10. Pazianas, M., S. Epstein & M. Zaidi. 2009. Evaluating the antifracture efficacy of bisphosphonates. *Rev. Recent Clin. Trials* **4:** 122–130.

11. Bussard, K.M., C.V. Gay & A.M. Mastro. 2008. The bone microenvironment in metastasis; what is special about bone? *Cancer Metastasis Rev.* **27:** 41–55.

12. Reeder, J.G. & A.M. Brufsky. 2010. The role of bisphosphonates in the adjuvant setting for breast cancer. *Oncology (Williston Park)* **24:** 462–467, 475.

13. Maughan, K.L., M.A. Lutterbie & P.S. Ham. 2010. Treatment of breast cancer. *Am. Fam. Physician* **81:** 1339–1346.

14. Brown, S.A. & T.A. Guise. 2009. Cancer treatment-related bone disease. *Crit. Rev. Eukaryot. Gene Expr.* **19:** 47–60.

15. Gnant, M. *et al.* 2009. Endocrine therapy plus zoledronic acid in premenopausal breast cancer. *N. Engl. J. Med.* **360:** 679–691.

16. Mauri, D. *et al.* 2010. Does adjuvant bisphosphonate in early breast cancer modify the natural course of the disease? A meta-analysis of randomized controlled trials. *J. Natl. Compr. Cancer Netw.* **8:** 279–286.

17. Lattouf, J.B. & F. Saad. 2010. Bone complications of androgen deprivation therapy: screening, prevention, and treatment. *Curr. Opin. Urol.* **20:** 247–252.

18. Russell, R.G. *et al.* 2007. Bisphosphonates: an update on mechanisms of action and how these relate to clinical efficacy. *Ann. N.Y. Acad. Sci.* **1117:** 209–257.

19. Asahi, H. *et al.* 2006. Bisphosphonate induces apoptosis and inhibits pro-osteoclastic gene expression in prostate cancer cells. *Int. J. Urol.* **13:** 593–600.

20. Lipton, A. 2010. Should bisphosphonates be utilized in the adjuvant setting for breast cancer? *Breast Cancer Res. Treat.* **122:** 627–636.

21. Borutaite, V. 2010. Mitochondria as decision-makers in cell death. *Environ. Mol. Mutagen* **51:** 406–416.

22. Braun, S. *et al.* 2005. A pooled analysis of bone marrow micrometastasis in breast cancer. *N. Engl. J. Med.* **353:** 793–802.

23. Aft, R. *et al.* 2010. Effect of zoledronic acid on disseminated tumour cells in women with locally advanced breast cancer: an open label, randomised, phase 2 trial. *Lancet Oncol.* **11:** 421–428.

24. Rack, B. *et al.* 2010. Effect of zoledronate on persisting isolated tumour cells in patients with early breast cancer. *Anticancer Res.* **30:** 1807–1813.

25. Braun, S. *et al.* 1999. Monoclonal antibody therapy with edrecolomab in breast cancer patients: monitoring of elimination of disseminated cytokeratin-positive tumor cells in bone marrow. *Clin. Cancer Res.* **5:** 3999–4004.

26. Bozionellou, V. *et al.* 2004. Trastuzumab administration can effectively target chemotherapy-resistant cytokeratin-19 messenger RNA-positive tumor cells in the peripheral blood and bone marrow of patients with breast cancer. *Clin. Cancer Res.* **10:** 8185–8194.

27. Verdijk, R., H.R. Franke, F. Wolbers & I. Vermes. 2007. Differential effects of bisphosphonates on breast cancer cell lines. *Cancer Lett.* **246:** 308–312.

28. Fromigue, O., L. Lagneaux & J.J. Body. 2000. Bisphosphonates induce breast cancer cell death in vitro. *J. Bone Miner. Res.* **15:** 2211–2221.

29. Diel, I.J. *et al.* 1998. Reduction in new metastases in breast cancer with adjuvant clodronate treatment. *N. Engl. J. Med.* **339:** 357–363.

30. Powles, T. *et al.* 2002. Randomized, placebo-controlled trial of clodronate in patients with primary operable breast cancer. *J. Clin. Oncol.* **20:** 3219–3224.

31. Diel, I.J. *et al.* 2008. Adjuvant oral clodronate improves the overall survival of primary breast cancer patients with micrometastases to the bone marrow: a long-term follow-up. *Ann. Oncol.* **19:** 2007–2011.

32. Eidtmann, H. *et al.* 2010. Efficacy of zoledronic acid in postmenopausal women with early breast cancer receiving adjuvant letrozole: 36-month results of the ZO-FAST study. *Ann. Oncol.* [Epub ahead of print].

33. Pantel, K., R.J. Cote & O. Fodstad. 1999. Detection and clinical importance of micrometastatic disease. *J. Natl. Cancer Inst.* **91:** 1113–1124.

34. Filleul, O., E. Crompot & S. Saussez. 2010. Bisphosphonate-induced osteonecrosis of the jaw: a review of 2,400 patient cases. *J. Cancer Res. Clin. Oncol.* **136:** 1117–1124.

35. Kabelitz, D., D. Wesch & W. He. 2007. Perspectives of gammadelta T cells in tumor immunology. *Cancer Res.* **67:** 5–8.

36. Gober, H.J. *et al.* 2003. Human T cell receptor gammadelta cells recognize endogenous mevalonate metabolites in tumor cells. *J. Exp. Med.* **197:** 163–168.

37. Raikkonen, J. *et al.* 2009. Zoledronic acid induces formation of a pro-apoptotic ATP analogue and isopentenyl pyrophosphate in osteoclasts in vivo and in MCF-7 cells in vitro. *Br. J. Pharmacol.* **157:** 427–435.

38. Monkkonen, H. *et al.* 2007. Zoledronic acid-induced IPP/ApppI production in vivo. *Life Sci.* **81:** 1066–1070.

39. Pettit, A.R. *et al.* 2001. TRANCE/RANKL knockout mice are protected from bone erosion in a serum transfer model of arthritis. *Am. J. Pathol.* **159:** 1689–1699.

40. Korganow, A.S. *et al.* 1999. From systemic T cell self-reactivity to organ-specific autoimmune disease via immunoglobulins. *Immunity* **10:** 451–461.

41. Francis, M.D., K. Hovancik & R.W. Boyce. 1989. NE-58095: a diphosphonate which prevents bone erosion and preserves joint architecture in experimental arthritis. *Int. J. Tissue React.* **11:** 239–252.

42. Eggelmeijer, F. *et al.* 1996. Increased bone mass with pamidronate treatment in rheumatoid arthritis. Results of a three-year randomized, double-blind trial. *Arthritis Rheum.* **39:** 396–402.

Ann. N.Y. Acad. Sci. ISSN 0077-8923

ANNALS OF THE NEW YORK ACADEMY OF SCIENCES
Issue: *Molecular and Integrative Physiology of the Musculoskeletal System*

# Molecular physiology of cardiac regeneration

Paola Bolli and Hina W. Chaudhry

Cardiovascular Regenerative Medicine, Cardiovascular Institute, Mount Sinai School of Medicine, New York, New York

Address for correspondence: Hina W. Chaudhry, M.D., Cardiovascular Regenerative Medicine, Cardiovascular Institute, Mount Sinai School of Medicine, New York, NY 10029. hina.chaudhry@mssm.edu

Heart disease is the leading cause of death in the industrialized world. This is partially attributed to the inability of cardiomyocytes to divide in a significant manner, and therefore the heart responds to injury through scar formation. One of the challenges of modern medicine is to develop novel therapeutic strategies to facilitate regeneration of cardiac muscle in the diseased heart. Numerous methods have been studied and a wide variety of cell types have been considered. To date, bone marrow stem cells, endogenous populations of cardiac stem cells, embryonic stem cells, and induced pluripotent stem cells have been investigated for their ability to regenerate infarcted myocardium, although stem cell transplantation has produced ambiguous results in human clinical trials. Several studies support another approach that seems very appealing: enhancing the limited endogenous regenerative capacity of the heart. The recent advances in stem cell and regenerative biology are giving rise to the view that cardiac regeneration, although not quite ready for clinical treatment, may translate into therapeutic reality in the not too distant future.

Keywords: cardiac; physiology; stem cell; regeneration; heart disease

## Introduction

The fascination with regenerative biology can be traced back through several millennia. In ancient Greek mythology, Prometheus was able to regenerate his liver on a daily basis despite having it attacked in a recurring fashion by an eagle sent by Zeus. Currently, heart disease is the leading cause of death worldwide. This is, in part, attributed to a limited regenerative capacity. Because adult mammalian cardiomyocytes lack significant replicative potential, cardiac muscle principally heals through formation of scar tissue after injury. As a result, contractile function declines, and patients can progress to heart failure.[1]

The challenge of cardiovascular regenerative medicine is to develop novel therapeutic strategies to facilitate regeneration of normally functioning cardiac muscle in the diseased heart. To this end, many investigators have explored a myriad of approaches in the last decade.

Regeneration of the infarcted heart implies creation of new myocardium, with myocytes that are electromechanically coupled to the host tissue and an appropriate coronary vasculature and connective tissue to support function.[2] Recent advances in stem cell and regenerative biology are giving rise to the view that cardiac regeneration, although not quite ready for prime time in terms of clinical treatment, is becoming an achievable reality. To date, numerous methods have been studied and a wide variety of cell types have been considered as candidates to attain this goal, but the results of stem cell transplantation are somewhat ambiguous and the ideal cell type has not yet been established. The use of bone marrow cells to regenerate infarcted myocardium has been investigated in hundreds of studies since the initial publication in the field almost 10 years ago.[3] Currently, however, a consensus is emerging that the ability of bone marrow-derived stem cells to differentiate into cardiomyocytes is questionable. Less controversy surrounds the evidence from several groups demonstrating that endogenous populations of cardiac stem cells have replicative and potentially regenerative capacities. Mammalian myocardium is now known to include a small population of stem-like cells expressing one of the cell surface markers c-kit, Sca-1, ABCG2, or one of the transcription

doi: 10.1111/j.1749-6632.2010.05814.x

factors, Islet-1 or nkx2.5.[4–8] Recent studies support another approach that seems very appealing: enhancing the limited endogenous regenerative capacity of the heart. Although mammalian cardiomyocytes do not retain the ability to divide in any significant manner, studies have shown that manipulation of cellular proteins may be able to promote cell cycle reentry and proliferation of cardiomyocytes.[9–16] In this review, we discuss the current knowledge of molecular physiology of cardiac regeneration in the mammalian heart.

## Studies of cardiac regeneration

A powerful approach to understanding the potential for human cardiac regeneration is to gain insights into cardiac regeneration in lower organisms (Fig. 1). Regenerative ability is retained in a widespread manner in metazoan phylogeny but has been lost in many species for reasons that are not understood.[17] Urodele amphibians have retained an extraordinary capacity to replace lost anatomical structures through a process known as epimorphic regeneration. An adult urodele, such as the newt or axolotl, is capable of regenerating its limbs, spinal cord, jaws, lens, retinas, and large sections of the heart.[17] Newt regeneration hinges on the local plasticity of differentiated cells that remain after injury or tissue removal, rather than the existence of a reserve-cell mechanism. This involves reentry to the cell cycle and loss of differentiated characteristics, so as to generate a local progenitor cell of restricted potentiality.[17]

Poss et al.[18,19] reported that 60 days after the removal of the cardiac apex, the zebrafish heart appeared to have completely regenerated. They demonstrated that the regeneration resulted from proliferation of cardiomyocytes adjacent to the area of injury. They also showed that regeneration failed to occur in fish harboring a mutation in mps1, a gene encoding a mitotic checkpoint kinase required for cell proliferation and regeneration of the zebrafish fin. This implies that a single gene affecting cardiomyocyte mitosis can impede the regenerative process.

By contrast, in mammalian hearts, cardiomyocytes rarely divide after injury. However, Piero Anversa and colleagues[11] challenged this existing dogma in 2001 by demonstrating that the adult heart has a subpopulation of myocytes that are not terminally differentiated and that are able to reenter the

cell cycle and undergo nuclear mitotic division early after myocardial infarction (MI). They reported that the number of cycling myocytes was 0.08% in the infarct border zone and 0.03% in the distant myocardium, and that the multiplication of myocytes was markedly attenuated as the length of time after MI increased. They had used Ki-67 expression to identify nuclei that had entered the cell cycle and sarcomeric α-actin to localize these nuclei to cardiomyocytes. Ki-67 labels all phases of the cell cycle except for $G_0$ and therefore is not a definitive marker for mitosis.[20] Nevertheless, this study opened a new wave of investigation into this complex and clinically critical area of research.

More recently, an elegant study by Bergmann and colleagues[21] examined the rate of cardiomyocyte turnover based on the observation that nuclear weapons tested in earth's atmosphere through 1963 had labeled all of the cells of the entire world's population. They demonstrated that cardiomyocytes regenerate but at a very slow rate, renewing with a gradual decrease from a 1% turnover annually at the age of 25 to 0.45% at the age of 75, and that fewer than 50% of cardiomyocytes are exchanged during a normal life span.

Therefore, this process is clearly inadequate in countering the massive myocyte loss seen after MI, whereby up to a billion cardiomyocytes can be lost.[22] Investigators worldwide are thus pursuing a variety of angles to enhance the regenerative process in the damaged heart, and the focus has largely been on stem and progenitor cells for treatment and prevention of heart failure. We have summarized the variety of stem and progenitor cell approaches to regenerate diseased myocardium below.

### Bone marrow–derived cells

The question of whether bone marrow–derived cells in general, and hematopoietic stem cells in particular, can transdifferentiate into cardiomyocytes and promote widespread regeneration has been hotly debated for years.[2,23,24]

Bone marrow–derived hematopoietic stem cells (BMCs) have been thought to exhibit the potential to differentiate into cardiomyocytes following transplantation.[3,25,26] However, three later studies[23,24,27] rigorously challenged the conclusions of these reports by independently demonstrating that BMCs transplanted into damaged hearts could not give rise to cardiomyocytes. Balsam et al.[24]

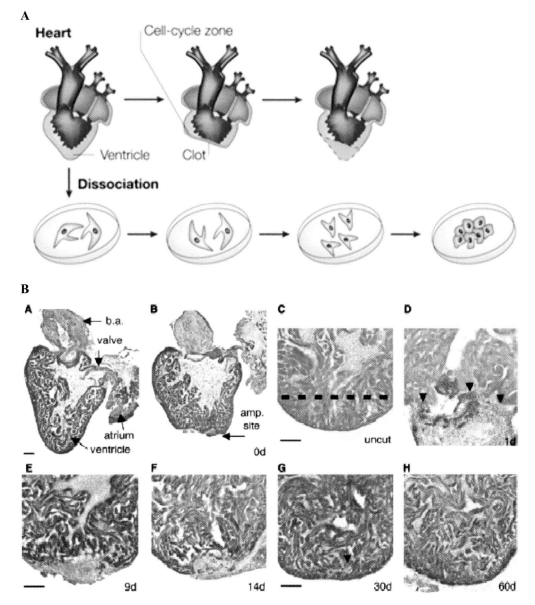

**Figure 1.** Although mammalian cardiomyocytes have not retained the ability to regenerate, cardiac regeneration is retained in lower organisms. (A) When newt cardiomyocytes are dispersed in culture, they reenter the cell cycle, proliferate, and re-aggregate to form a beating syncytium. From Brockes and Kumar, *Nature Reviews*, 2002. (B) Regeneration of ventricular myocardium in the resected zebrafish heart. Sixty days after the removal of the cardiac apex, the zebrafish heart appeared to have completely regenerated. From Poss *et al.*[19]

have shown that BMCs not only fail to give rise to cardiomyocytes, but also actually develop into different blood cell types, despite being in the heart. The beneficial effects noted in earlier studies in terms of ventricular performance might be partially attributable to angioblast-mediated vasculogenesis,[28] which could prevent apoptosis

of native cardiomyocytes rather than by direct myogenesis.

R. Bolli and colleagues[29] have noted that the adult bone marrow harbors a population of small CXCR4$^+$ cells that are nonhematopoietic and express markers of lineage commitment for several different tissues. These very small embryonic-like stem cells (VSELs) are $Sca1^+/Lin^-/CD45^-$ and express cardiac markers, including Nkx2.5/Csx, GATA-4, and MEF2C, and acquire the cardiomyocytic phenotype *in vitro* under specific culture conditions. The authors then investigated the role of these bone marrow–derived VSELs after MI in a mouse model. They reported that after MI, VSELs-treated mice exhibited improved global and regional left ventricular systolic function and attenuated myocyte hypertrophy in surviving tissue compared with controls. Moreover, they showed that pluripotent VSELs are mobilized from the bone marrow into the peripheral blood after an acute MI, with a peak in the circulating levels early after MI and followed by decrease at seven days.[29,30]

Some have explored the use of mesenchymal stem cells (MSCs) found in the adult bone marrow. Although a few investigators support the idea that these cells could undergo myogenic differentiation,[31] Dzau and colleagues[32–34] demonstrated that the cytoprotective effects on cardiomyocytes from genetically modified mesenchymal cells were largely attributable to paracrine effects and not to *de novo* regeneration by cardiomyogenic differentiation.

### Nonmarrow-derived mesenchymal cells

The regenerative potential of human bone marrow-derived mesenchymal cells is limited due to low efficiency for cardiomyogenic transdifferentiation. Therefore, nonmarrow-derived mesenchymal cells have also been studied given the premise that these cells may have higher cardiomyogenic transdifferentiation efficiency. To date, menstrual blood–derived mesenchymal cells,[35] umbilical cord blood–derived MSCs,[36] and placental chorionic blood–derived mesenchymal cells[37] have been examined. Given that these cells would be used as allografts, this would likely lead to problems of immunologic rejection. Recently, amniotic membrane-derived mesenchymal stem cells (AMCs) were reported to have potential for transdifferentiation into cells of various organs. Tsuji *et al.*[38] showed that human AMCs (hAMCs) can be transdifferentiated into cardiomyocytes *in vitro* and *in vivo*. The cardiomyogenic transdifferentiation efficiency of hAMCs was significantly higher than that of marrow-derived MSCs. Furthermore, xenografted hAMCs transdifferentiated into cardiomyocytes and survived more than two weeks, thus suggesting that hAMCs were tolerated *in situ*.

However, in an elegant editorial that accompanies this report, Penn and Mayorga surmise that the underlying reason for improvement in cardiac function is the recruitment of hibernating cells to generate contractile work, rather than significant regeneration of cardiomyocytes. In fact, the "new" cardiomyocytes derived from the hAMCs were scattered throughout the infarct border zone, intercalated in between native cardiomyocytes that survived the infarct.[39]

### Endogenous cardiac stem cells

Given the limitations of BMCs, the search for naturally occurring, authentic heart progenitor cells continues in earnest, with several groups having reported on the existence of such cells.[4,6,16,40–42]

**Side population cells.** Side population (SP) cells have stem cell characteristics, as they have been shown to contribute to diverse lineages. Interest in this population relates to findings showing that SP cells can serve as progenitors for hematopoietic cells,[43] skeletal muscle,[44] and endothelium.[25] SP cells have been identified in the bone marrow as well as in nonhematopoietic tissues, including skeletal muscle,[45] mammary gland,[46] heart,[4] liver, brain, kidney, and lung,[43,47] based on their capability to efflux fluorescent vital dyes like Hoechst 33342.[48] One study demonstrated that as few as 2,000–5,000 SP cells isolated from adult bone marrow were able to reconstitute the irradiated *mdx* mouse bone marrow.[44] Later these cells were recruited from the bone marrow to participate in skeletal muscle repair. In another study, as few as 100 skeletal muscle SP cells were shown to reconstitute the entire bone marrow of a lethally irradiated mouse.[25] SP cells have a unique ability to extrude fluorescent vital dye Hoechst 33342, which is readily taken up by live cells where it binds to DNA.[48] Analysis of these cells on a flow cytometer equipped with an ultraviolet (UV) laser source permits detection of these cells. When unpurified murine bone marrow cells labeled with Hoechst are examined by fluorescence-activated cell sorter analysis, SP cells fall within a separate population to the side of the remaining cells on a dot plot of emission data, hence the term "side population." SP cells have been identified in several species, including mice, rhesus monkeys, swine, and humans.[49–51]

The ability of SP cells to efflux Hoechst 33342 is dependent on the expression of the Abcg2 (also known as Bcrp1: breast cancer resistance protein) protein, an ATP-binding cassette transporter.[52] Abcg2 is a member of the family of ABC transporters and was first identified in a breast cancer cell line.[53] Abcg2 has been described as a cause of the multidrug resistance phenotype. Downregulation in Abcg2 expression is seen in various committed hematopoietic progenitors. The expression of Abcg2, however, is not limited to hematopoietic stem cells, as its expression is found in SP cells from diverse sources, including monkey bone marrow, mouse skeletal muscle, bone marrow, and embryonic stem (ES) cells.[47]

Garry and his group[4] have demonstrated that Abcg2 is expressed in cardiac progenitor cells, and the highest levels of expression during cardiac development are noted at embryonic day (E) 8.5. They have further illustrated that the adult heart contains Abcg2-expressing SP cells which are capable of proliferation and differentiation, and that these cells are capable of participating in myocardial repair after cryoinjury is induced in the mouse heart.[54]

In 2007, Komuro and colleagues[55] reported that cardiac SP cells from postnatal rat heart differentiate into cardiomyocytes both *in vitro* and *in vivo*. In this study, oxytocin and trichostatin A induced postnatal cardiac SP cells to differentiate into beating cardiomyocytes. After intravenous transplantation of SP cells in normal adult rats, SP cells migrated and homed in the interstitial space of myocardium. When SP cells were intravenously transplanted into the cryoinjured heart, the number of SP cells was significantly larger in the border zone than in the remote or infarct zones after transplantation. Liao and her colleagues have shown that CD31⁻/Sca1⁺ cardiac SP cells are capable of both biochemical and functional cardiomyogenic differentiation into mature cardiomyocytes, with expression of cardiomyocyte-specific transcription factors and contractile proteins, as well as stimulated cellular contraction and intracellular calcium transients indistinguishable from adult cardiomyocytes. Moreover, they illustrated the necessity of cell-extrinsic signaling through coupling, although not fusion, with adult cardiomyocytes in regulating cardiomyogenic differentiation of cardiac SP cells.[56]

**Sca-1⁺ cells.** Schneider and colleagues[6] reported the existence of adult heart–derived cardiac progenitor cells expressing stem cell antigen-1 (Sca-1) in 2003. They illustrated that these cells express CD31 but are distinct from hematopoietic stem cells (based on CD45, CD34, c-kit, Lmo2, GATA2, and Tal1/Scl), endothelial progenitors cells (based on CD45, CD34, Flk-1, and Flt-1), muscle "satellite" cells (Sca-1⁻, CD34⁺), muscle-derived hematopoietic stem cells (CD45⁺), and multipotential muscle cells (Sca-1⁺, CD34⁺). In response to 5-azacytidine, these cells differentiated into cells expressing cardiac-specific markers *in vitro* and, when given intravenously after ischemia/reperfusion, Sca-1⁺ cells homed to injured myocardium.[6] However, fusion of transplanted cells with native cardiomyocytes was noted in addition to transdifferentiation, and the cycling cardiomyocytes were derived from transplanted Sca-1⁺ cells, and not the endogenous ones. There is no clear evidence that sustained functional benefit is obtained with transplanted Sca-1⁺ cells.

**c-Kit⁺ cells.** Beltrami *et al.*[41] reported the existence of a population of Lin⁻ c-Kit⁺ cells in the adult rat heart that, *in vivo* and *in vitro*, exhibit all the properties expected for cardiac stem cells. These cells are uniformly negative for markers of hematopoietic lineages and proteins specific for myocytes, endothelial, smooth muscle cells or fibroblasts, but partially expressed cardiac transcription factors Nkx2.5, GATA4, and MEF2. The c-Kit⁺ cells were self-renewing, clonogenic, and multipotent. This cardiac progenitor population has been reported to be able to differentiate into myogenic, endothelial or smooth muscle cell lineages *in vitro*; however, it was not mentioned whether the c-Kit⁺ cells formed functional, beating cardiomyocytes *in vitro*. When injected into an ischemic heart, these cells formed viable new blood vessels and myocytes with the characteristic of young cells, reportedly encompassing up to 70% of the ventricle. Dawn *et al.*[57] have shown that these cells are able to regenerate infarcted myocardium and improve heart function when delivered via intravascular route. However, cells from human biopsy samples that expressed c-Kit were reported to coexpress markers of mast cells and to lack expression of cardiac transcription factors NKX2–5 and Islet 1,[58] crucial markers of the cardiac progenitor cell state in fetal hearts.[8,59]

**Islet-1⁺ cells.** During vertebrate cardiac development, the structure of the heart is formed by the differentiation and interaction of multiple fields.

The primary and secondary heart fields of the embryonic disc give rise to the intracardiac structures of the heart under the influence of adjacent tissues. The secondary heart field, which contributes to two-thirds of the myocardium, has been considered to be well delineated by the expression of Islet-1 (Isl1), a member of the LIM homeodomain transcription family.[60] This cardiac progenitor population has been first identified in postnatal rat, mouse, and human myocardium[42,61] and reportedly does not extrude Hoechst 33342 and does not express c-kit and Sca-1 markers, thus identified as a distinct population of progenitors.[42,60,61]

Recently, Moretti et al.[59] reported that Isl1[+] cells isolated from 1- to 5-day old mouse pups are multipotent and able to differentiate into cardiac, smooth muscle, and endothelial cells. The role these cells can potentially play in cardiac repair after injury has not been described.

**Epicardial progenitor cells.** Evans and her colleagues[62] recently reported a previously unknown myocardial progenitor derived from *Tbx18*-expressing epicardial cells. The authors illustrated that these progenitor cells migrate into the outer cardiac surface to form the epicardium, and then make a substantial contribution to myocytes in the ventricular septum, atrium, and ventricular walls. These cardiac progenitors cells also give rise to cardiac fibroblasts and coronary smooth muscle cells. Previous studies had already demonstrated that the proepicardium and/or epicardium is a source for coronary vascular progenitors and cardiac fibroblast. Moreover, the regeneration of the zebrafish heart has been shown to be associated with reactivation of this early marker of proepicardium/ epicardium, the T-box transcription factor *Tbx18*.[63]

An additional epicardial population of cardiac progenitors has been identified by Pu and colleagues.[64] These cardiomyocyte progenitors are marked by the expression of the transcription factor Wt1. The authors showed that, during murine heart development, a subset of these Wt1[+] precursors differentiated into fully functional cardiomyocytes. Wt1[+] cells arose from progenitors that express Nkx2–5 and Isl1 suggesting that they share a developmental origin with multipotent Nkx2–5 and Isl1 progenitors. It has not been demonstrated whether epicardial progenitor cell types can induce cardiac repair after injury.

## ES cells

ES cells meet all the requirements for stem cells: clonality, self-renewal, and multipotentiality. ES cells are derived from the inner cell mass of mammalian blastocysts and have the ability to grow indefinitely while maintaining pluripotency and the ability to differentiate into cells of all three germ layers.[65] ES cells have the potential to completely regenerate the myocardium. ES cells have been shown to be able to differentiate into cardiomyocytes that exhibit spontaneous contractile activity and express cardiac transcription factors.[66,67] Moreover, ES cells differentiate into immature cardiomyocytes when transplanted in injured heart, and can regenerate infarcted myocardium resulting in an improvement in cardiac structure and function, although not in a sustained manner over time.[68,69]

Despite promising results with ES cells, there are ethical issues regarding the use of embryonic material, the problem of immunological reactions, as well as the tendency of ES cells to form teratomas. There is also the concern that ES cells could, in the course of differentiation, proliferate in an uncontrolled way *in vivo* as shown in the case of neuronal progenitors.[70] Therefore, clinical use of these cells is still fraught with obstacles.

## Induced pluripotency and nuclear reprogramming

Adult cells are genetically equivalent to early ES. Differential gene expression is the result of reversible epigenetic changes that are gradually imposed on the genome during development. The reversal of the differentiation state of the mature cell to one that is characteristic of the undifferentiated embryonic state is defined as nuclear reprogramming. Three approaches have been successfully used to induce nuclear reprogramming: nuclear transfer, cell fusion, and transcription factor transduction (Fig. 2).[71]

A remarkable breakthrough in this field was made in 2006 when Takahashi and Yamanaka[72] discovered that by reintroducing only four factors (Oct 3/4, Sox2, c-Myc, and Klf4), somatic cells can be reprogrammed into "induced" pluripotent stem cells (iPS). Subsequent refinement of this technique resulted in the generation of iPS cells with as few as two factors from mouse fibroblasts.[73,74] Recent studies have demonstrated the ability to achieve cardiomyocyte differentiation of iPS cells,[75,76] but the reprogramming is incomplete

## A) Nuclear transfer

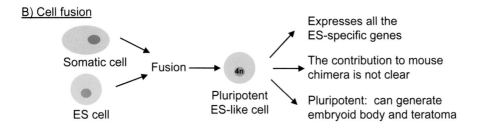

## B) Cell fusion

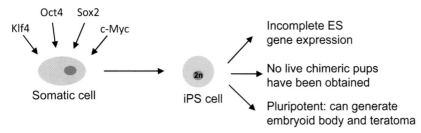

## C) Transcription-factor transduction

**Figure 2.** Methods of nuclear reprogramming. (A) In nuclear transfer, the nucleus of a somatic cell is transplanted into an enucleated oocyte and used to produce mouse ES cells that are totally reprogrammed. (B) In cell fusion, a somatic cell is fused with an ES cell, leading to the reprogramming of the somatic cell's nucleus. (C) Transcription-factor transduction: the introduction of four genes (Oct3/4, Sox2, c-Myc, and Klf4) by using retroviral vectors can generate iPS cells.

and therefore clinical and pharmacological use of iPS cells still needs further clarification. Although therapeutic use of iPS-derived cardiomyocytes is unclear, the diagnostic use of this technology is beginning to be realized. Lemischka and colleagues[77] generated iPS cells from patients with LEOPARD syndrome, an autosomal-dominant developmental disorder that shows hypertrophic cardiomyopathy as major phenotype. The authors established iPS cell lines from two LEOPARD syndrome patients and reported that these iPS cells have the ability to differentiate *in vitro* and *in vivo* into cell types representative of all three germ layers and into a variety of haematopoietic cell types, including early haematopoietic progenitors (CD41[+]), early erythroblast (CD71[+]/CD235a[+]), and macrophages (CD11b[+]). They also observed that *in vitro*–derived cardiomyocytes from LEOPARD syndrome iPS cells had significantly increased median surface area compared to wild-type iPS cardiomyocytes. Moreover, the LEOPARD syndrome cardiomyocytes had a higher degree of sarcomeric organization and preferential localization of NFATC4 (regulator of cardiac hypertrophy) in the nucleus.

The ability to reprogram endogenous cardiac fibroblasts into cardiomyocytes could avoid the reprogramming to iPS before cardiac differentiation. Very recently, Srivastava's group[78] has demonstrated that the combination of three transcription factors—Gata4, Mef2c, and Tbx5—can rapidly and efficiently induce cardiomyocytes-like cells from postnatal cardiac and dermal fibroblasts. The induced cardiomyocytes were reported to be similar to neonatal cardiomyocytes in global gene expression profile, electrophysiologic parameters, and could contract simultaneously. Although much refinement and characterization is necessary, these findings raise the possibility of reprogramming endogenous fibroblasts into functional cardiomyocytes for regenerative purposes.

## Studies in dedifferentiation

Terminally differentiated mammalian cells are normally incapable of reversing the differentiation process. By contrast, urodele amphibians such as the newt are capable of remarkable plasticity in cellular differentiation.[17] During limb regeneration, cells underlying the wound epithelium dedifferentiate to form a pool of proliferating, progenitor cells known as the blastema.[79] These cells later redifferentiate and a new limb develops. Muscle cells are known to participate in this dedifferentiation process contributing to the formation of cartilage.[80,81]

A nuclear protein that is thought to play a role in urodele cellular dedifferentiation is the homeobox-containing transcriptional repressor *msx1*. This protein is expressed in the early regeneration blastema[82] and its expression in the developing mouse limb demarcates the boundary between the undifferentiated (*msx1*-expressing) and differentiating (no *msx1* expression) cells.[82] Keating and colleagues[83] have shown that when ectopic *msx1* was expressed in mouse C2C12 myotubes, there was a reduction in the levels of muscle differentiation proteins such as MyoD, myogenin, MRF4, and p21. A subset of these myotubes cleaved to produce a pool of proliferating mononucleated cells. Furthermore, clonal populations of the myotube-derived mononucleate cells could be induced to redifferentiate into cells expressing chondrogenic, adipogenic, myogenic, and osteogenic markers depending on which growth medium they were placed in. These results suggest that terminally differentiated mammalian myotubes do retain the ability to dedifferentiate when stimulated with the appropriate signals and that *msx1* can contribute to the dedifferentiation process.

With respect to the potential for adult cardiomyocytes to dedifferentiate, studies of the *in vitro* behavior of newt cardiomyocytes demonstrate that a subgroup of these cells has not undergone the complete program of terminal differentiation, and, in addition, the absence of signals keeps these cells quiescent in a noninjured animal.[84] Approximately 29% of newt ventricular embryonic cardiomyocytes placed in culture progressed through one or more rounds of cell division (including karyokinesis and cytokinesis) giving rise to beating mononucleate cells. The peak of mitotic activity was noted at 10 days associated with myofibril disassembly and reassembly.[85]

Kuhn *et al.*[12] illustrated that periostin, a component of the extracellular matrix, when delivered through the cardiac cellular matrix, induced reentry of differentiated mononucleated cardiomyocytes into the cell cycle in rats. After experimental MI, periostin reduced infarct size and fibrosis and improved cardiac function suggesting that this component of the extracellular matrix may enhance the regenerative capacity of adult mammalian hearts.

A more recent study demonstrated that the growth factor neuregulin1-(NRG1) induced mononuclear cardiomyocytes to divide, and that NRG1-induced cardiomyocyte proliferation requires the NRG1 tyrosine kinase receptor, ErbB4 (v-erb-a erythroblastic leukemia viral oncogene homolog 4). Furthermore, mononucleated differentiated cardiomyocytes disassembled their sarcomeres during division, and NRG1-induced cardiomyocyte proliferation promoted cardiac repair.[14]

Thus, the potential to stimulate further dedifferentiation of adult mammalian cardiomyocytes warrants additional clarification of mechanisms involved.

## The cell cycle and cardiomyocytes

The potential to reactivate cardiomyocyte proliferation through manipulation of putative cell cycle regulators offers an exciting impetus for the design of novel therapeutic interventions to enhance cardiac function during disease conditions. This approach has received considerable interest because of the identification of key cell cycle regulatory proteins and several reports suggesting that manipulation of these factors can reactivate cell cycle activity in postnatal cardiomyocytes or myocardium.[10,15,86–89] Progression through the cell cycle is regulated by cyclins complexed with their catalytic subunits known as cyclin-dependent kinases (Cdks). Cyclin A2 complexed with Cdk2 is essential for the G1/S transition and cyclin A2/Cdk1 promotes entry into mitosis[90] (Fig. 3). Other cyclins, regulating the passage from G1 to S phase, include D-type cyclins and cyclin

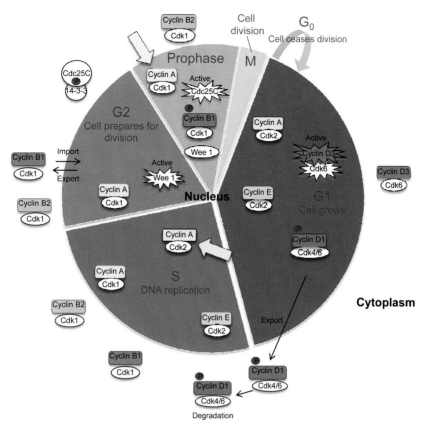

**Figure 3.** Mammalian cell cycle. Progression through the cell cycle is regulated by cyclins complexed with their catalytic subunits, known as cyclin-dependent kinases (Cdks). Cyclin A2 is unique among the cyclins in that it is critical for both major transitions of the cell cycle, G1/S and G2/M.

E.[91,92] B-type cyclins are associated with Cdk1 to control entry into and exit from mitosis.

Cyclin A2 is absolutely essential for normal embryonic development to occur. A targeted deletion of cyclin A2 in the mouse exhibited embryonic lethality at embryonic day 5.5.[93] Cyclin A2 is the only cyclin to be completely silenced after birth at both the message and the protein level in rat, human,[94] and mouse cardiogenesis,[15] coincident with withdrawal of cardiomyocytes from the cell cycle.[94] It is also the only cyclin to regulate the two major transitions of the cell cycle, G1/S and G2/M.[91,92] We have shown, through the use of a transgenic mouse model with constitutive expression of cyclin A2, that direct manipulation of the cell cycle can be achieved and that mitosis can be induced in differentiated cardiomyocytes.[9,13,15,16] In the uninjured transgenic mouse, the increase in cardiomyocyte mitoses is most notable during early postnatal development, slowing down a few weeks after birth

as the transgene protein product is predominantly found in the cytoplasm as the mouse approaches adulthood. When these mice are subjected to MI, induced by ligation of the left anterior artery, they exhibit enhanced cardiac function in a serial and sustained manner over 3 months with a concomitant reduction in myocardial damage as assessed by MRI when compared to nontransgenic littermates.[16] This response appears to involve reentry of the periinfarct myocardial cells into the cell cycle as demonstrated by an increase in mitotic index and DNA synthesis. Furthermore, mitosis is noted in small ($\sim$5 $\mu$m) cells found in the infarct zone that express markers of cardiomyocytes as well as ABCG2, a known marker of the SP phenotype[52] with predominantly nuclear localization of the cyclin A2 protein. These small cells appear to represent immature cardiomyocytes. These data indicate that cardiomyocytes derived from ABCG2-expressing progenitors in the infarcted myocardium recapitulate the developmental paradigm noted in the early postnatal cyclin A2 transgenic hearts;[15] that is, mitosis is potentiated in postnatal cardiomyocytes-expressing cyclin A2. Thus, it appears that cyclin A2 is directing increased rounds of mitosis of the "immature" cardiomyocytes in the infarct zones of the transgenic mouse hearts. Additionally, we noted cell cycle reentry of periinfarct myocardium, which indicates a retention of perhaps a more "plastic" phenotype in the transgenic heart, reminiscent of that noted in urodele amphibians such as the newt.[84,85]

Furthermore, postnatal (immature) transgenic cardiomyocytes in culture exhibit a significantly higher mitotic index than nontransgenic cells and even undergo cytokinesis, thus reinforcing the role of cyclin A2 in cardiac repair. Even after four weeks in culture, transgenic cardiomyocytes had formed a beating syncytium,[16] a phenomenon that has never previously been associated with mammalian cardiomyocytes to our knowledge. We have more recently noted that when SP cells are isolated from transgenic mouse hearts, they undergo significantly more mitoses and exhibit greater efficiency of differentiation into cardiomyocytes than compared to SP cells isolated from nontransgenic hearts (unpublished data).

One of the main limitations noted in the studies using endogenous cardiac progenitors and ES cells for transplantation into infarcted hearts has been the lack of sustained functional improvement of cardiac function over time. Transplantation is also plagued by the need to assure electromechanical coupling of the exogenously supplied cells with the native cardiomyocytes. It appears that the use of native progenitors will comprise an important goal in cardiac regenerative medicine. However, these native progenitors are, at their naturally occurring prevalence and levels of activation, clearly inadequate in reversing the downward spiral of events culminating in heart failure. Their differentiation in response to environmental cues might be expected to generate cardiomyocytes of a postmitotic nature, hence limiting the ability of such endogenous processes to counter the massive myocyte death in MI. A proliferative stimulus, such as that provided by cyclin A2, may provide the "missing link" in being able to effectively spur endogenous progenitor cells to repopulate the damaged heart.

In support of the utility of cyclin A2 in cardiac repair, we have demonstrated that when cDNA

encoding cyclin A2 is delivered via adenovirus to the periinfarct zone of infarcted hearts in genetically naive adult rats, cardiomyocyte regeneration is noted in the infarct border zone and heart failure is averted when compared to controls receiving empty adenovirus.[9,13] More recently, we have studied this process in a large animal model and administered adenoviral cyclin A2 to infarcted porcine hearts (Fig. 4). When compared to controls receiving a null adenovirus, the experimental animals exhibit significantly enhanced cardiac function as measured by both two-dimensional and three-dimensional echocardiography and MRI.[95] Preliminary histological data indicate enhanced mitoses in the periinfarct zones of the pigs that had received the cyclin A2 adenovirus and an increased frequency of stem cell homing.

In conclusion, our studies demonstrated that cyclin A2, normally silenced in the postnatal mammalian heart, has regenerative capacity. Additionally, we found that the therapeutic delivery of cyclin A2 induces cardiomyocyte cell cycle activation with significant enhancement of cardiac function in small and large animals, thus providing the strongest proof-of-concept to date that cell cycle manipulation offers potent therapeutic benefits in heart disease.

## Clinical trials

Although the efficiency of bone marrow–derived stem cells for cardiac repair is still controversial, the success of earlier studies indicating that BMCs might induce cardiac repair prompted a flurry of clinical trials. The largest trials done to date are the REPAIR-AMI (Reinfusion of Enriched Progenitor Cells and Infarct Remodeling in Acute Myocardial Infarction) and the ASTAMI (Autologous Stem Cell Transplantation in Acute Myocardial Infarction) trials.[96,97] In the REPAIR-AMI trial, intracoronary administration of BMCs in patients with acute MI was associated with slightly improved recovery of left ventricular contractile function. After one year, infusion of BMCs was associated with a reduction in the prespecified combined clinical end point of death, recurrence of MI, and any revascularization procedure. In contrast, the ASTAMI trial showed no effects of intracoronary injection of autologus mononuclear BMCs on global left ventricular function.

Although the early returns from clinical trials suggest that the delivery of autologous BMCs to

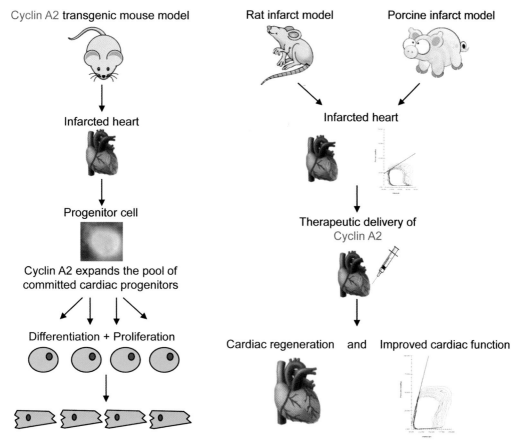

**Figure 4.** Cyclin A2 transgenic mice exhibit cardiac regeneration and improvement in cardiac function after MI. In the infarct and border zones, cyclin A2 increases cycling of cardiac progenitors. When delivered in a therapeutic manner via adenovirus, cyclin A2 also mediates cardiac repair in the rat and porcine infarct models with evidence of cardiomyocyte regeneration in the border zones, scar reduction, and improvement of left ventricular ejection fraction (LVEF).

diseased hearts may have marginal benefit, fortunately they have not proven to be unsafe. The greatest benefit noted thus far, a 3% increase in ejection fraction in a patient subset of the REPAIR-AMI trial, may be due to a paracrine effect that is yet to be defined. A firm conclusion about the ability of bone marrow cells to promote true tissue regeneration cannot be drawn at this time. Further investigation into cell types with true cardiac transdifferentiation capacity is necessary in order to realize the prospects of cell therapy in human patients.

## Future goals

The success of cardiac regenerative therapy is a critical clinical imperative, with implications that can alter the disease profile of humankind. Ultimately, interdisciplinary efforts will be needed to assure clinical reality. A greater understanding is required

of the cell types than can undergo most efficient and effective cardiac differentiation, the cues for migration to the sites of injury, how electromechanical coupling of transplanted cell types can be achieved, and how cell death of transplanted progenitors can be mitigated. Because recent evidence has emerged that the adult mammalian heart has a limited degree of cardiomyocyte turnover, the administration of extracellular factors to stimulate the endogenous turnover of cardiomyocytes or proliferation of endogenous progenitors may represent the best approach, or may perhaps be combined with cell-based therapy.

## Conflicts of interest

Hina W. Chaudhry is founder of VentriNova, Inc., which seeks to develop clinical therapies based on work with cyclin A2. H.W. Chaudhry is also an inventor of an awarded patent and other pending patents concerning this technology.

## References

1. Malki, Q. *et al*. 2002. Clinical presentation, hospital length of stay, and readmission rate in patients with heart failure with preserved and decreased left ventricular systolic function. *Clin. Cardiol.* **25:** 149–152.
2. Murry, C.E. *et al*. 2004. Haematopoietic stem cells do not transdifferentiate into cardiac myocytes in myocardial infarcts. *Nature* **428:** 664–668.
3. Orlic, D. *et al*. 2001. Bone marrow cells regenerate infarcted myocardium. *Nature* **410:** 701–705.
4. Martin, C.M. *et al*. 2004. Persistent expression of the ATP-binding cassette transporter, Abcg2, identifies cardiac SP cells in the developing and adult heart. *Dev. Biol.* **265:** 262–275.
5. Beltrami, A.P. *et al*. 2001. Cardiac c-kit positive cells proliferate in vitro and generate new myocardium in vivo. *Circulation* **104:** 324–324.
6. Oh, H. *et al*. 2003. Cardiac progenitor cells from adult myocardium: homing, differentiation, and fusion after infarction. *Proc. Natl. Acad. Sci. USA* **100:** 12313–12318.
7. Nakano, A., H. Nakano & K.R. Chien. 2008. Multipotent islet-1 cardiovascular progenitors in development and disease. *Cold Spring Harb. Symp. Quant. Biol.* **73:** 297–306.
8. Wu, S.M. *et al*. 2006. Developmental origin of a bipotential myocardial and smooth muscle cell precursor in the mammalian heart. *Cell* **127:** 1137–1150.
9. Woo, Y.J. *et al*. 2006. Therapeutic delivery of cyclin A2 induces myocardial regeneration and enhances cardiac function in ischemic heart failure. *Circulation* **114:** I206–I213.
10. Pasumarthi, K.B., H. Nakajima, H.O. Nakajima, *et al*. 2005. Targeted expression of cyclin D2 results in cardiomyocyte DNA synthesis and infarct regression in transgenic mice. *Circ. Res.* **96:** 110–118.
11. Beltrami, A.P. *et al*. 2001. Evidence that human cardiac myocytes divide after myocardial infarction. *N. Engl. J. Med.* **344:** 1750–1757.
12. Kuhn, B. *et al*. 2007. Periostin induces proliferation of differentiated cardiomyocytes and promotes cardiac repair. *Nat. Med.* **13:** 962–969.
13. Woo, Y.J. *et al*. 2007. Myocardial regeneration therapy for ischemic cardiomyopathy with cyclin A2. *J. Thorac. Cardiovasc. Surg.* **133:** 927–933.
14. Bersell, K., S. Arab, B. Haring & B. Kuhn. 2009. Neuregulin1/ErbB4 signaling induces cardiomyocyte proliferation and repair of heart injury. *Cell* **138:** 257–270.
15. Chaudhry, H.W. *et al*. 2004. Cyclin A2 mediates cardiomyocyte mitosis in the postmitotic myocardium. *J. Biol. Chem.* **279:** 35858–35866.
16. Cheng, R.K. *et al*. 2007. Cyclin A2 induces cardiac regeneration after myocardial infarction and prevents heart failure. *Circ. Res.* **100:** 1741–1748.
17. Brockes, J.P. & A. Kumar. 2002. Plasticity and reprogramming of differentiated cells in amphibian regeneration. *Nat. Rev. Mol. Cell Biol.* **3:** 566–574.
18. Poss, K.D., A. Nechiporuk, A.M. Hillam, *et al*. 2002. Mps1 defines a proximal blastemal proliferative compartment essential for zebrafish fin regeneration. *Development* **129:** 5141–5149.
19. Poss, K.D., L.G. Wilson & M.T. Keating. 2002. Heart regeneration in zebrafish. *Science* **298:** 2188–2190.
20. Scholzen, T. & J. Gerdes. 2000. The Ki-67 protein: from the known and the unknown. *J. Cell. Physiol.* **182:** 311–322.
21. Bergmann, O. *et al*. 2009. Evidence for cardiomyocyte renewal in humans. *Science* **324:** 98–102.
22. Reinecke, H., E. Minami, W.Z. Zhu & M.A. Laflamme. 2008. Cardiogenic differentiation and transdifferentiation of progenitor cells. *Circ. Res.* **103:** 1058–1071.
23. Nygren, J.M. *et al*. 2004. Bone marrow-derived hematopoietic cells generate cardiomyocytes at a low frequency through cell fusion, but not transdifferentiation. *Nat. Med.* **10:** 494–501.
24. Balsam, L.B. *et al*. 2004. Haematopoietic stem cells adopt mature haematopoietic fates in ischaemic myocardium. *Nature* **428:** 668–673.
25. Jackson, K.A. *et al*. 2001. Regeneration of ischemic cardiac muscle and vascular endothelium by adult stem cells. *J. Clin. Invest.* **107:** 1395–1402.
26. Orlic, D. *et al*. 2001. Mobilized bone marrow cells repair the infarcted heart, improving function and survival. *Proc. Natl. Acad. Sci. USA* **98:** 10344–10349.
27. Murry, C.E. *et al*. 2004. Haematopoietic stem cells do not transdifferentiate into cardiac myocytes in myocardial infarcts. *Nature* **428:** 664–668.
28. Kocher, A.A. *et al*. 2001. Neovascularization of ischemic myocardium by human bone-marrow-derived angioblasts prevents cardiomyocyte apoptosis, reduces remodeling and improves cardiac function. *Nat. Med.* **7:** 430–436.
29. Dawn, B. *et al*. 2008. Transplantation of bone marrow-derived very small embryonic-like stem cells attenuates left ventricular dysfunction and remodeling after myocardial infarction. *Stem Cells* **26:** 1646–1655.
30. Zuba-Surma, E.K. *et al*. 2007. Transplantation of bone marrow-derived very small embryonic-like stem cells (VSELs) improves left ventricular function and remodeling after myocardial infarction. *Circulation* **116:** 204–204.
31. Toma, C., M.F. Pittenger, K.S. Cahill, *et al*. 2002. Human mesenchymal stem cells differentiate to a cardiomyocyte phenotype in the adult murine heart. *Circulation* **105:** 93–98.
32. Gnecchi, M. *et al*. 2006. Evidence supporting paracrine hypothesis for Akt-modified mesenchymal stem cell-mediated cardiac protection and functional improvement. *Faseb Journal* **20:** 661–669.
33. Mangi, A.A. *et al*. 2003. Mesenchymal stem cells modified with Akt prevent remodeling and restore performance of infarcted hearts. *Nat. Med.* **9:** 1195–1201.
34. Gnecchi, M. *et al*. 2005. Paracrine action accounts for marked protection of ischemic heart by Akt-modified mesenchymal stem cells. *Nat. Med.* **11:** 367–368.
35. Hida, N. *et al*. 2008. Novel cardiac precursor-like cells from human menstrual blood-derived mesenchymal cells. *Stem Cells* **26:** 1695–1704.
36. Nishiyama, N. *et al*. 2007. The significant cardiomyogenic potential of human umbilical cord blood-derived mesenchymal stem cells in vitro. *Stem Cells* **25:** 2017–2024.

37. Okamoto, K. *et al*. 2007. 'Working' cardiomyocytes exhibiting plateau action potentials from human placenta-derived extraembryonic mesodermal cells. *Exp. Cell Res.* **313:** 2550–2562.

38. Tsuji, H. *et al*. 2010. Xenografted human amniotic membrane-derived mesenchymal stem cells are immunologically tolerated and transdifferentiated into cardiomyocytes. *Circ. Res.* **106:** 1613–1623.

39. Penn, M.S. & M.E. Mayorga. 2010. Searching for understanding with the cellular lining of life. *Circ. Res.* **106:** 1554–1556.

40. Matsuura, K. *et al*. 2004. Adult cardiac Sca-1-positive cells differentiate into beating cardiomyocytes. *J. Biol. Chem.* **279:** 11384–11391.

41. Beltrami, A.P. *et al*. 2003. Adult cardiac stem cells are multipotent and support myocardial regeneration. *Cell* **114:** 763–776.

42. Laugwitz, K.L. *et al*. 2005. Postnatal isl1+ cardioblasts enter fully differentiated cardiomyocyte lineages. *Nature* **433:** 647–653.

43. Asakura, A. & M.A. Rudnicki. 2002. Side population cells from diverse adult tissues are capable of in vitro hematopoietic differentiation. *Exp. Hematol.* **30:** 1339–1345.

44. Gussoni, E. *et al*. 1999. Dystrophin expression in the mdx mouse restored by stem cell transplantation. *Nature* **401:** 390–394.

45. Asakura, A., P. Seale, A. Girgis-Gabardo & M.A. Rudnicki. 2002. Myogenic specification of side population cells in skeletal muscle. *J. Cell Biol.* **159:** 123–134.

46. Welm, B.E. *et al*. 2002. Sca-1(pos) cells in the mouse mammary gland represent an enriched progenitor cell population. *Dev. Biol.* **245:** 42–56.

47. Summer, R. *et al*. 2003. Side population cells and Bcrp1 expression in lung. *Am. J. Physiol. Lung Cell. Mol. Physiol.* **285:** L97–L104.

48. Goodell, M.A., K. Brose, G. Paradis, *et al*. 1996. Isolation and functional properties of murine hematopoietic stem cells that are replicating in vivo. *J. Exp. Med.* **183:** 1797–1806.

49. Goodell, M.A. *et al*. 1997. Dye efflux studies suggest that hematopoietic stem cells expressing low or undetectable levels of CD34 antigen exist in multiple species. *Nat. Med.* **3:** 1337–1345.

50. Storms, R.W., M.A. Goodell, A. Fisher, *et al*. 2000. Hoechst dye efflux reveals a novel CD7(+)CD34(−) lymphoid progenitor in human umbilical cord blood. *Blood* **96:** 2125–2133.

51. Uchida, N., T. Fujisaki, A.C. Eaves & C.J. Eaves. 2001. Transplantable hematopoietic stem cells in human fetal liver have a CD34(+) side population (SP)phenotype. *J. Clin. Invest.* **108:** 1071–1077.

52. Zhou, S. *et al*. 2001. The ABC transporter Bcrp1/ABCG2 is expressed in a wide variety of stem cells and is a molecular determinant of the side-population phenotype. *Nat. Med.* **7:** 1028–1034.

53. Doyle, L.A. *et al*. 1998. A multidrug resistance transporter from human MCF-7 breast cancer cells. *Proc. Natl. Acad. Sci. USA* **95:** 15665–15670.

54. Martin, C.e.a. 2004. Abcg2 cardiac stem cell population partecipates in repair of adult mouse and human heart. *Suppl. Circ.* **110** [Abstract no. 811].

55. Oyama, T. *et al*. 2007. Cardiac side population cells have a potential to migrate and differentiate into cardiomyocytes in vitro and in vivo. *J. Cell Biol.* **176:** 329–341.

56. Pfister, O. *et al*. 2005. CD31(−) but not CD31(+) cardiac side population cells exhibit functional cardiomyogenic differentiation. *Circ. Res.* **97:** 52–61.

57. Dawn, B. *et al*. 2005. Cardiac stem cells delivered intravascularly traverse the vessel barrier, regenerate infarcted myocardium, and improve cardiac function. *Proc. Natl. Acad. Sci. USA* **102:** 3766–3771.

58. Pouly, J. *et al*. 2008. Cardiac stem cells in the real world. *J. Thorac. Cardiovasc. Surg.* **135:** 673–678.

59. Moretti, A. *et al*. 2006. Multipotent embryonic isl1+ progenitor cells lead to cardiac, smooth muscle, and endothelial cell diversification. *Cell* **127:** 1151–1165.

60. Cai, C.L. *et al*. 2003. Isl1 identifies a cardiac progenitor population that proliferates prior to differentiation and contributes a majority of cells to the heart. *Dev. Cell* **5:** 877–889.

61. Bu, L. *et al*. 2009. Human ISL1 heart progenitors generate diverse multipotent cardiovascular cell lineages. *Nature* **460:** U113–U130.

62. Cai, C.L. *et al*. 2008. A myocardial lineage derives from Tbx18 epicardial cells. *Nature* **454:** 104–108.

63. Lepilina, A. *et al*. 2006. A dynamic epicardial injury response supports progenitor cell activity during zebrafish heart regeneration. *Cell* **127:** 607–619.

64. Zhou, B. *et al*. 2008. Epicardial progenitors contribute to the cardiomyocyte lineage in the developing heart. *Nature* **454:** 109–113.

65. Thomson, J.A. *et al*. 1998. Embryonic stem cell lines derived from human blastocysts. *Science* **282:** 1145–1147.

66. Xu, C., S. Police, N. Rao & M.K. Carpenter. 2002. Characterization and enrichment of cardiomyocytes derived from human embryonic stem cells. *Circ. Res.* **91:** 501–508.

67. Yang, L. *et al*. 2008. Human cardiovascular progenitor cells develop from a KDR+ embryonic-stem-cell-derived population. *Nature* **453:** 524–528.

68. van Laake, L.W. *et al*. 2007. Human embryonic stem cell-derived cardiomyocytes survive and mature in the mouse heart and transiently improve function after myocardial infarction. *Stem Cell Res.* **1:** 9–24.

69. Nussbaum, J. *et al*. 2007. Transplantation of undifferentiated murine embryonic stem cells in the heart: teratoma formation and immune response. *FASEB J.* **21:** 1345–1357.

70. Roy, N.S. *et al*. 2006. Functional engraftment of human ES cell-derived dopaminergic neurons enriched by coculture with telomerase-immortalized midbrain astrocytes. *Nat. Med.* **12:** 1259–1268.

71. Hochedlinger, K. & R. Jaenisch. 2006. Nuclear reprogramming and pluripotency. *Nature* **441:** 1061–1067.

72. Takahashi, K. & S. Yamanaka. 2006. Induction of pluripotent stem cells from mouse embryonic and adult fibroblast cultures by defined factors. *Cell* **126:** 663–676.

73. Kim, J.B. *et al*. 2008. Pluripotent stem cells induced from adult neural stem cells by reprogramming with two factors. *Nature* **454:** 646–650.

74. Liu, Y.H., R. Karra & S.M. Wu. 2008. Cardiovascular stem cells in regenerative medicine: ready for prime time? *Drug Discov. Today Ther. Strateg.* **5:** 201–207.

75. Freund, C., R.P. Davis, K. Gkatzis, *et al.* 2010. The first reported generation of human induced pluripotent stem cells (iPS cells) and iPS cell-derived cardiomyocytes in the Netherlands. *Neth. Heart J.* **18:** 51–54.

76. Zhang, J.H. *et al.* 2009. Functional cardiomyocytes derived from human induced pluripotent stem cells. *Circ. Res.* **104:** E30–E41.

77. Carvajal-Vergara, X. *et al.* 2010. Patient-specific induced pluripotent stem-cell-derived models of LEOPARD syndrome. *Nature* **465:** U808–U812.

78. Ieda, M. *et al.* 2010. Direct reprogramming of fibroblasts into functional cardiomyocytes by defined factors. *Cell* **142:** 375–386.

79. Brockes, J.P. 1997. Amphibian limb regeneration: rebuilding a complex structure. *Science* **276:** 81–87.

80. Lo, D.C., F. Allen & J.P. Brockes. 1993. Reversal of muscle differentiation during urodele limb regeneration. *Proc. Natl. Acad. Sci. USA* **90:** 7230–7234.

81. Kumar, A., C.P. Velloso, Y. Imokawa & J.P. Brockes 2000. Plasticity of retrovirus-labelled myotubes in the newt limb regeneration blastema. *Dev. Biol.* **218:** 125–136.

82. Simon, H.G. *et al.* 1995. Differential expression of myogenic regulatory genes and Msx-1 during dedifferentiation and redifferentiation of regenerating amphibian limbs. *Dev. Dyn.* **202:** 1–12.

83. Odelberg, S.J., A. Kollhoff & M.T. Keating. 2000. Dedifferentiation of mammalian myotubes induced by msx1. *Cell* **103:** 1099–1109.

84. Bettencourt-Dias, M., S. Mittnacht & J.P. Brockes 2003. Heterogeneous proliferative potential in regenerative adult newt cardiomyocytes. *J. Cell. Sci.* **116:** 4001–4009.

85. Kaneko, H., M. Okamoto & K. Goshima. 1984. Structural change of myofibrils during mitosis of newt embryonic myocardial cells in culture. *Exp. Cell. Res.* **153:** 483–498.

86. Poolman, R.A., J.M. Li, B. Durand & G. Brooks. 1999. Altered expression of cell cycle proteins and prolonged duration of cardiac myocyte hyperplasia in p27KIP1 knockout mice. *Circ. Res.* **85:** 117–127.

87. Agah, R. *et al.* 1997. Adenoviral delivery of E2F-1 directs cell cycle reentry and p53-independent apoptosis in postmitotic adult myocardium in vivo. *J. Clin. Invest.* **100:** 2722–2728.

88. Jackson, T. *et al.* 1990. The c-myc proto-oncogene regulates cardiac development in transgenic mice. *Mol. Cell Biol.* **10:** 3709–3716.

89. Soonpaa, M.H. *et al.* 1997. Cyclin D1 overexpression promotes cardiomyocyte DNA synthesis and multinucleation in transgenic mice. *J. Clin. Invest.* **99:** 2644–2654.

90. Sherr, C.J. & J.M. Roberts. 1995. Inhibitors of mammalian G1 cyclin-dependent kinases. *Genes Dev.* **9:** 1149–1163.

91. Weinberg, R.A. 1995. The retinoblastoma protein and cell cycle control. *Cell* **81:** 323–330.

92. Hunter, T. 1997. Oncoprotein networks. *Cell* **88:** 333–346.

93. Murphy, M. *et al.* 1997. Delayed early embryonic lethality following disruption of the murine cyclin A2 gene. *Nat. Genet.* **15:** 83–86.

94. Yoshizumi, M. *et al.* 1995. Disappearance of cyclin A correlates with permanent withdrawal of cardiomyocytes from the cell cycle in human and rat hearts. *J. Clin. Invest.* **95:** 2275–2280.

95. Shapiro, S., Y. Kawase, D. Ladage, M.G. Guzman, *et al.* 2009. Administration of cyclin A2 by gene transfer improves cardiac function in a porcine model of myocardial infarction. *Circulation* **120:** S754.

96. Lunde, K. *et al.* 2006. Intracoronary injection of mononuclear bone marrow cells in acute myocardial infarction. *N. Engl. J. Med.* **355:** 1199–1209.

97. Schachinger, V. *et al.* 2006. Intracoronary bone marrow-derived progenitor cells in acute myocardial infarction. *N. Engl. J. Med.* **355:** 1210–1221.